Farm Equipment
and
HAND TOOLS.

Farm Equipment

and

HAND TOOLS.

A Practical Manual.

EDITED BY

GEORGE A. MARTIN.

THE STEPHEN GREENE PRESS

BRATTLEBORO, VERMONT

This book has been produced in the United States of America. It is published by The Stephen Greene Press, Brattleboro, Vermont 05301.

LIBRARY OF CONGRESS CATALOGING IN PUBLICATION DATA

Martin, George A d. 1904.
Farm equipment and hand tools.
Reprint of the 1887 ed. published by O. Judd Co., New York under title: Farm appliances, with a new foreword and index.
Bibliography: p.
Includes index.
1. Farm equipment. 2. Agricultural implements.
I. Title.
S675.M285 1980 631.3 79-27074
ISBN 0-8289-0356-5

TABLE OF CONTENTS.

PUBLISHER'S NOTE.

Farm Equipment and Hand Tools: A Practical Manual was first published in 1887 with the title *Farm Appliances: A Practical Manual.* The original publisher was the Orange Judd Company of New York. Orange Judd was one of the most active publishers in our history. Through the last third of the nineteenth century, and well into the twentieth, the company brought out scores of practical books on all aspects of agriculture and husbandry, and also had large lines of books on dogs, horses, fishing and hunting, boating, natural history and architecture.

The present edition of *Farm Equipment and Hand Tools* includes a foreword by Grant Heilman, America's foremost agricultural photographer-writer. Mr. Heilman's photographs of rural and agricultural subjects have appeared in every major American publication using photographic illustration. He also writes and lectures widely on agricultural and conservation topics. His books include *Farm Town* and *Wheat Country,* both published by The Stephen Greene Press.

In addition to the foreword the publisher has added to the present edition of this book a selected bibliography and an index. Otherwise the volume is substantially a facsimile of the 1887 edition.

FOREWORD.

When I was a kid I occasionally was allowed to spend time at Aunt Myra and Uncle Dory's chicken farm across the Delaware down in southern New Jersey. Uncle Dory was a great admirer of clever thinking, possibly because he didn't do much of it. When he came across a stimulating idea he always called it a "brain whetter." I was in college before I realized he was saying "whetter," not "wetter."

That's what this book is, a real "brain whetter." Uncle Dory would have loved it. Sitting on the old stump out beside the egg collecting shed, he would have slapped his knee in appreciation, and muttered: "These are real brain whettin' ideas."

Written almost a hundred years ago, *Farm Equipment and Hand Tools* will still find an extensive, varied, and appreciative audience.

For those who are traveling back to the land as a way of life, sections of the book are a practical guide to some of the problems of living off the land. Not everyone, for example, knows how to hang, clean, and scald a hog—and if you have to do it and don't know how, you've got a problem. I know, I've tried it. Back on page 188 you'll find an even half-dozen ways to scald a hog, all sensible. And, just for good measure, if you want to smoke the resulting meat, you'll find directions for a "cheap temporary smoke-house."

But more than just a practical sort of how-to book, *Farm Equipment and Hand Tools* is, in Uncle Dory's parlance, a real "brain whetter" in other ways as well. It puts forth ideas that not only solve practical farm problems (try: five different ways to hang a bag for filling) but also stimulate creative thinking. You may find solutions to problems the author hadn't even thought of.

There is a chapter which carries a marvelous heading: "Stump-pullers, Derricks and Slings." It is really about the practical use of leverage. I took the same course in engineering school, where it was grandiosely called "Theoretical Mechanics," and was filled with outlandish mathematical formulae, and recourse to slide rules. The chapter has little by way of numbers, but it does tell all about how to pull big stumps, lift a whole bunch of loaded bags with an "endless rope sling," and make a "homemade horse-power."

But, completely aside from its practical applications, I found the book a real "brain whetter" just because it made me look at things as they were a hundred years back.

The editor insists, in his Introduction, that

> Inventive talent has completely revolutionized the processes of farming. The work which required the labor of many, under primitive methods, is now better done by one person with the aid of improved appliances . . .

That's one of those statements people drag out a hundred years later, publish without dates, then give you the dates and say: "Ha, ha, just shows how things are relative." They're right.

Around 1887, when the book was originally published by the Orange Judd Company, an offshoot of the venerable *American Agriculturist* magazine, more than half the families in the United States had to work on farms to support themselves and the non-farm population. Today it takes about one family in fifty to provide farm products for everyone in the United States, and to have enough left over to ship large quantities of farm products abroad—our largest source of foreign exchange.

While many of the appliances George Martin talks about and illustrates are still useful today (the fruit pickers he shows are still ingenious solutions to a tough problem), I was struck by some of the changes. Not necessarily the big, obvious changes, like productivity, but more subtle, and sometimes upsetting, changes.

One of the great differences, obviously, was the availability of craftsmen then who could make things. Martin is constantly referring to "a very simple contrivance which any machine shop can furnish"; or he talks of ". . . the fastenings (for a root washer) at each end of the barrel can be made by any blacksmith . . ."; or, "any pottery will furnish to order, or they may have them on hand, five or six gallon cylindrical jars of glazed ware, having a spigot hole in the side close to the bottom, and the usual jar lid . . ."

Somehow, I just don't know that my machine shop or my pottery could provide me with all that. And as for my blacksmith, well, he's always too busy patching up automobile innards to bother with fastenings for my root washer.

I'm not sure that such change is a good thing.

The farm equipment of George Martin's day seem to have about them a wholesomeness and a solidity which is difficult to come by today. He is casually accepting when he says:

> The frame (of a cheap home-made roller) is made of four by four oak, bolted together firmly. The logs are eighteen inches in diameter and three and one half feet long. . . .

Or, he says, ". . . an oak stick averaging half a foot in diameter, was selected from the wood pile, and a piece five foot long cut off . . ."

That upset me. Obviously Martin wasn't familiar with my hundred-year-later woodpile, which is mostly filled with scraps of plywood, abandoned because the knotholes on the finished side made them too rough to use.

And I envy his junk pile. Junk piles just aren't what they used to be, even on farms:

> An excellent portable-power can be made by taking a bevel-gearing from an old discarded brick-machine, a pair of spur wheels from an ancient reaper, two or three shafts and a band-wheel from other sources—all odds and ends picked up cheaply here and there . . .

I wasn't quite sure what a "portable-power" was, even after I read about it, but I felt sure it was made of marvelous things I could only dream of. Junk piles today seem to have mostly cracked automobile engine blocks, and what can you do with *them?*

Present-day craftsmen, looking for art objects to produce, could hardly do better than to reproduce in wood, carefully worked with a spokeshave and drawknife, the manger fork and chaff fork on page 67—they're just plain beautiful tools, and likely the wood needed can still be found.

But on the other hand, even though craftsmen and materials may have dwindled, Martin talks of the reduction in labor brought by his ingenious devices, but we recognize that farming then was still largely the application of large amounts of sweat. His one-paragraph description of digging a "Dakota Well" never mentions the weeks of back-breaking labor involved. The rig he sketches seems charming enough, but after the rig is set up, it's still all pick and shovel work. I'm afraid I'd opt for picking up the phone and calling Twin Lakes Drilling, and asking them to bring down one of their rotary rigs for a couple of hours.

We've lost the village smithy, even the local pottery, and the beauty of hand crafted tools. But we've gained *muscle:* the well driller with his towering rig, automatic hay balers, rat-tat-tatting chain saws, Steiger tractors. The most far reaching change in the past hundred years has been the substitution of oil power for horse and human sweat.

But this hasn't been free. It has meant increased interdependence. The farmer, or even the village smithy, is no longer capable of repairing a combine or a four wheel drive tractor. He's dependent on a specialist from the city, or from some factory on the other side of the globe.

And, while Martin's ingenious farmer was dependent on his horses, and horses had their problems, he was totally independent of the whims of oil producers of the Middle East. I've no thought that the shift from horses to diesel hasn't been worth it. But we may be forced to rethink some of our "progress," and Martin's book gives us a thoughtful look backward.

I've great confidence in the ingenuity of our farmers to overcome the technological problems of the future, including oil shortages. I just wish I would be able to look at the changes in the farm world over the *next* hundred years. Come to think of it, I wish George Martin could walk with me down through those next hundred years, chuckling at the ingenious solutions which our future farmers develop. Judging from his book, he'd have a wonderful time.

GRANT HEILMAN

INTRODUCTION.

Inventive talent has completely revolutionized the processes of farming. The work which required the labor of many, under primitive methods, is now better done by one person with the aid of improved appliances. To explain and illustrate some of the most practical and easily made appliances is the object of this volume. They are such as secure greater comfort to domestic animals, provide supplies of wholesome water, economize labor and assist in dispatching much of the important work on the farm. The hints and suggestions herein contained are the result of practical experience. It is believed that every farmer, gardener, householder, in fact every one interested in labor-saving contrivances, will find very much of interest and value in this volume.

George A. Martin

Farm Equipment
and
HAND TOOLS.

CHAPTER I.

RACKS, MANGERS, STANCHIONS AND TROUGHS.

RACKS AND FEED BOXES FOR HORSES.

There are various forms of racks, mangers and feed-boxes for horses. One of the worst devices is the old-

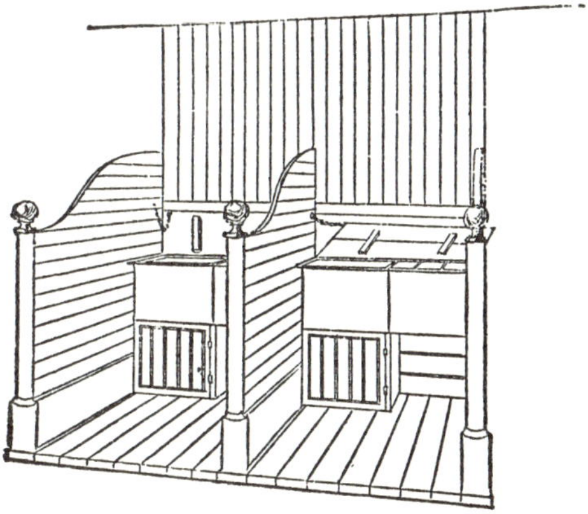

Fig. 1.—FEEDING RACK FOR HORSES.

fashioned hay-rack, extending from the manger high above the head of the horses, which are compelled to reach up for their hay. This is a most unnatural position for a horse, which does not, when out of the stable, take its food like a giraffe from trees, but from the ground. Aside from this, a high rack causes the double peril of getting dust into the lungs and other objects into the eyes of the horses. The above engraving shows an arrangement for hay and cut feed, or dry grain, which prevents waste, and is very convenient for the horse and its owner. The manger extends across the whole stall (a single one) and is reached through a falling door in the feeding passage. The hay box goes to the bottom, and has a barred door, through which the waste chaff may be removed, if it does not work out. The feed-box is protected by a barred cover, made of half-inch round iron, having spaces through which the horse can feed; but the bars prevent him from throwing out the feed or grain, in the attempt to pick out the best. The halter is run through a hole in the top of the manger, or a ring bolt in the side of the stall, and has a block of wood at the end, by the weight of which it is kept drawn tight, leaving no slack for the horse to get entangled with. When the horses are fed, the feeding door is shut and fastened by a button.

COVERED HORSE MANGER.

Horses will get their heads to the bottom of the hay

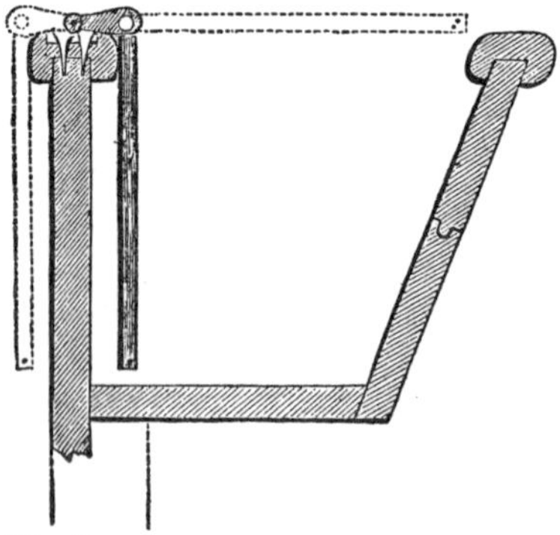

Fig. 2.—IMPROVED HORSE MANGER.

manger if they can, and will often throw the hay out, if not prevented. The illustration, figure 2, is taken from a stable, in which such annoyance is easily and simply prevented. A rack of iron rods, or of wood, is made and hinged to the top of the manger in front, so that it may be thrown up and over the front when the manger is filled, and then turned down upon the hay. The bars or rods are just far enough apart for the horse to get his nose through to the hay, but of course, he cannot get his head through. Iron is better than wood, because the horse cannot gnaw upon it. The bottom of every manger should be slatted, to let the hay seed and dust fall through—thus averting a frequent source of cough and heaves in horses.

FEEDING TROUGH AND HAY SHUTE.

To prevent waste of grain and hay, the trough and hay manger may be made as shown in the engravings here

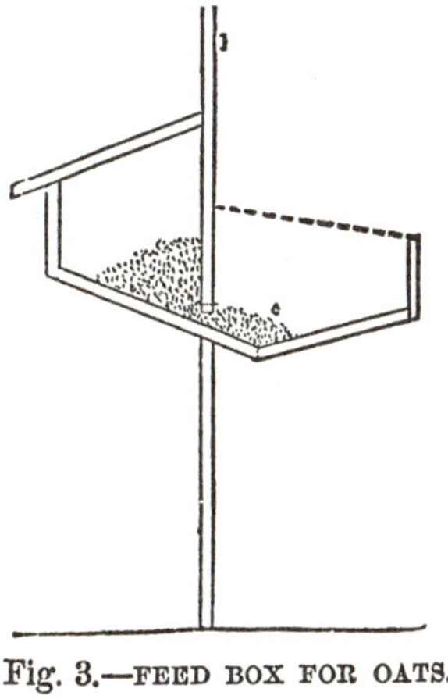

Fig. 3.—FEED BOX FOR OATS.

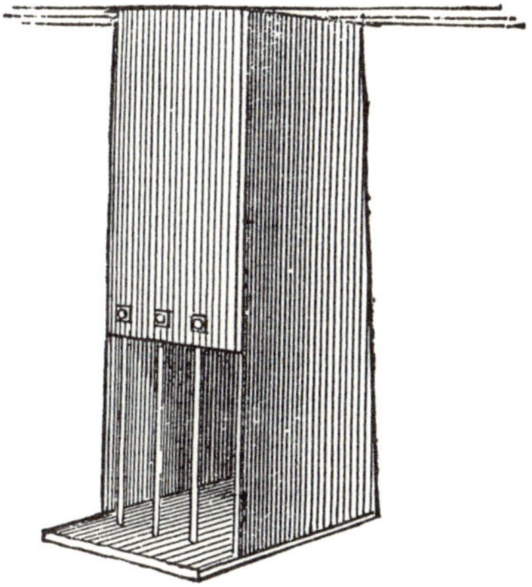

Fig. 4.—HAY FEED BOX.

given. The grain box (figure 3) is fixed in the front of the stall, a part of it projecting through the partition, into the feed passage, where there is a lifting hinged cover. The trough, of course, opens into the stall. In the center of the trough there is an upright division, open only for an inch or two, through which the grain or meal slides down little by little into the front division.

The hay shute is shown at figure 4. It comes from the floor above, where it has a hinged cover, which, if desired, is left open for ventilation. It increases in width downwards, to prevent the hay from lodging. The front is provided with small iron bars, to prevent the horse from pulling out the hay and thus causing loss. The bottom should be slatted, to allow the escape of dust.

DEVICE FOR BOX STALL.

For valuable animals it is best to have loose box-stalls. A range of such stalls can be built very cheaply, and as

Fig. 5.—FEED BOX AND HAY RACK IN BOX STALL.

the occupants need not be fastened, they can be quickly let out of the building in case of fire. The feeding arrangement for such stalls is shown at figure 5. It consists of a hay-rack in the corner, with a feed box near it. At the front of the feed box there is a falling door in the partition, through which, when it is half let down in a sloping position, the feed of grain, or the cut feed, may be placed in the box. The same arrangement may be used for the hay-rack, if the front is boarded up to the top; but if it is boarded only for five feet, the hay may be lifted over the top of it from the feeding passage.

FEED-BOX FOR EXTRA STALL.

There are times when the arrival of friends or other event calls for an extra stall. To provide for such emergencies, a feed box, and the way to use it, are shown in the engravings, figures 6 and 7. The trough, figure 6, is useful anywhere, it being a "fencetrough" or feed box. Upright pieces with mortises are made of inch stuff, and nailed on each side of the passage-way. Two by three-inch bars are used, entering into mortises on one side and dropping into slots on the other, the middle bar being keyed in. The upper bar is kept in

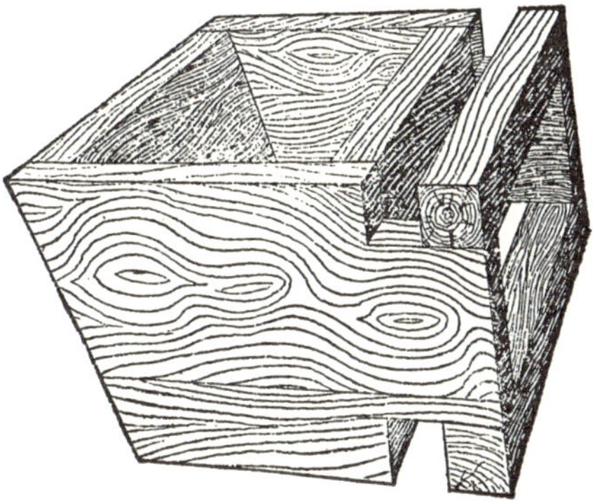

Fig. 6.—FENCE FEED BOX.

place by a swinging key put on the partition with a stout screw, and given a little play, so that it will drop by its own weight into its proper position. The feed box is made as in figure 6, with elongated sides extending through and beyond the rails or bars, with notches to receive the bars as indicated, made by nailing the pieces at the extreme ends across from side to side, as shown. The box being put on the lowest bar, close to the end of it, and the middle bar being placed in position and keyed, secures it. When the box is not in use, it is kept in the harness closet with the two lower bars. The top bar is

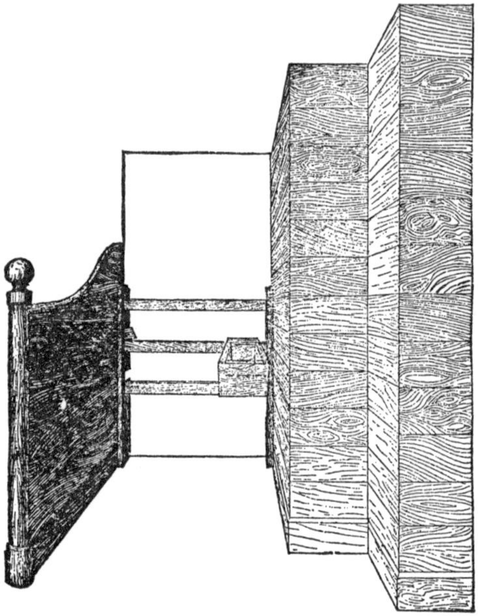

Fig. 7.—STALL IN PASSAGE-WAY.

generally left in place, to prevent horses, that might get loose, going into the carriage house.

VARIOUS CATTLE STANCHIONS.

In the engraving figure 8, one of the stanchions is shown open, and the other two closed. The pieces *d*, *e*, *f*, *g*, and *h*, are immovable, *a*, *b*, *c*, being the movable stanchions. The device consists of three strips, two inches wide, and three quarter inch thick, fastened to one upright piece by means of two bolts, *d* and *b*; the length of the strip is regulated by the distance between the stanchions. Bolts are also used at *a* and *c*, the bolt at *c* passing through a small block, two inches thick, which assists in moving the upright piece. A similar block, *e*, is also placed on the movable stanchion, upon which the block at *c* rests when the stanchion is closed.

The fastening *f*, and the piece *c*, are so arranged as to fall in place at the same time. It will be seen that the animal not only fastens herself in place, but she is doubly secured by the pieces *f* and *c*. (The block at *e* may be omitted if desired, and the device be used with the

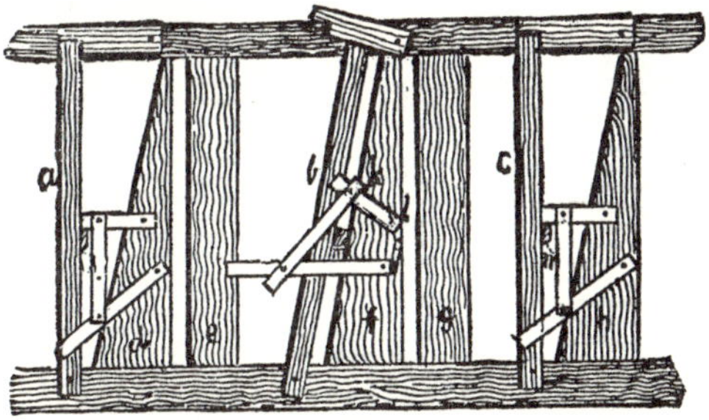

Fig. 8.—SELF-CLOSING CATTLE STANCHION.

fastening *f* only). A badly hooked cow is often the result of careless hired men, and such carelessness is obviated by the use of the above arrangement. A cow takes her place in the open stanchion, and in trying to get at the feed below, presses against the lever *a*, brings *c* to place, and closes the stanchion.

The engraving, figure 9, shows how every farmer who uses stanchions can arrange to close all the cows in at the same time. The two-inch strip *g*, is planed on all sides, and made to move easily in the loops *e*, *d*, which are of heavy galvanized iron, bent below so as to allow the strip to slide, and are attached to the immovable stanchions by screws. The hard wood pins *a*, *b*, *c*, extend about two inches through, so as to catch the movable stanchions. A lever is fixed at *h*, and attached to the movable strip. This device is comparatively inexpensive, and can be attached to all kinds of movable stanchions, generally used for fastening cows. Even after it is put on the stanchions, it need not be used unless desired. It has the advantage in being separate from every stanchion. One, two, or more animals may be closed in by hand and the balance with this device. It in no way interferes

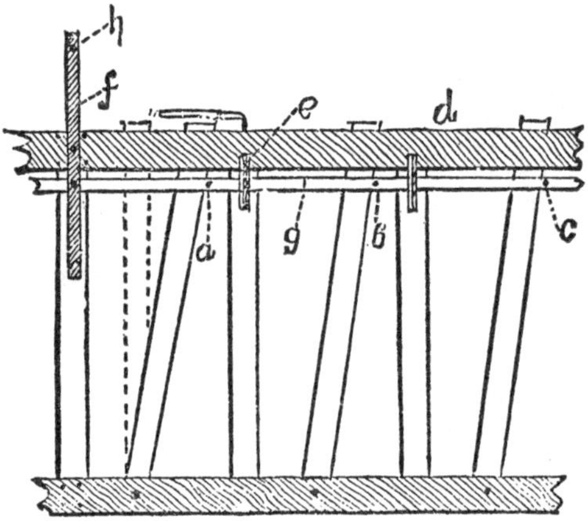

Fig. 9.—DEVICE FOR CLOSING CATTLE STANCHIONS.

with the necks of the cows, and saves a great many steps. If a person reaches over in front of the cows, to close them in by hand, he is in danger of being struck in the face with a horn. The above device removes this danger. It is simple and cheap.

The use of permanent neck-chains, locked on around the necks of breeding animals and young blooded stock, affords an excellent means of fastening the animals in their stalls. A chain and snap are attached to the stall, by which, the snap being caught into the ring of the neck chain, the animals are fastened. A better way is shown in the accompanying sketch of a cow stable. Two round stanchions are placed three feet apart for each stall, and are the only indications of subdivisions or stalls in the stable. A chain about eighteen inches long having a snap at one end, is attached by a ring to each stanchion. Both chains are made fast to the ring in the "necklace," and should have very little slack. If the stanchions are of hard wood, and smooth, the rings will slide easily up and down, but should not come within a

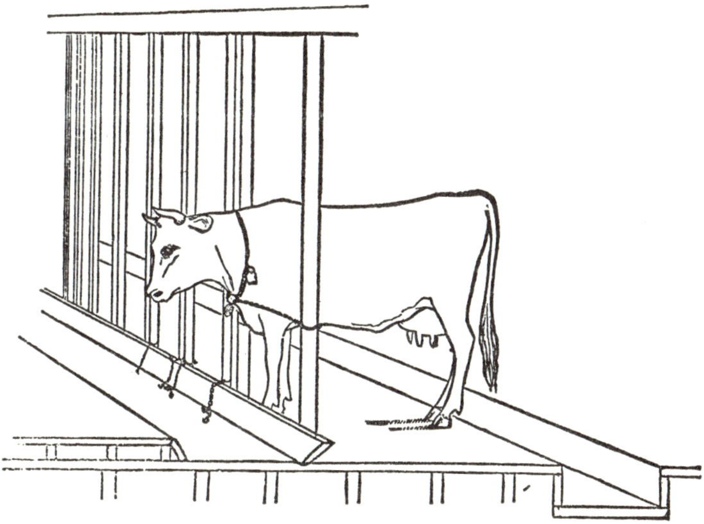

Fig. 10.—CHAIN CATTLE STANCHION.

foot of the floor. The cows will have free motion of the head to either side, can lie down and get up easily, but have very little motion forward and back, hence will keep on the platform and keep clean. They are besides kept perfectly devoted each to her own affairs, as she cannot reach over to either neighbor, to quarrel or to steal her forage.

The chief objection to the stanchion comes from its rigidity and vice-like grip, and any improvement in it should be in the direction of comfort to the animal, rather than in handier ways of fastening. The accompanying engraving shows how the rigid plan of the neck latches can be in part avoided. The greatest discomfort to stock, when stanchioned, comes when lying down. When standing, there is freedom of movement, but when the animal is down and attempts to rise, it is held fast. Stanchions made as here shown, avoid this. The neck-latches *a*, *b*, are not fastened at the bottom, but pass through the side block *c*, which rests on the lower stringer. By making this side block about eight inches shorter than the space between the uprights *d* and *c*, a swinging motion is obtained that gives considerable freedom. The bolt through the neck latch *a*, in the upper stringer,

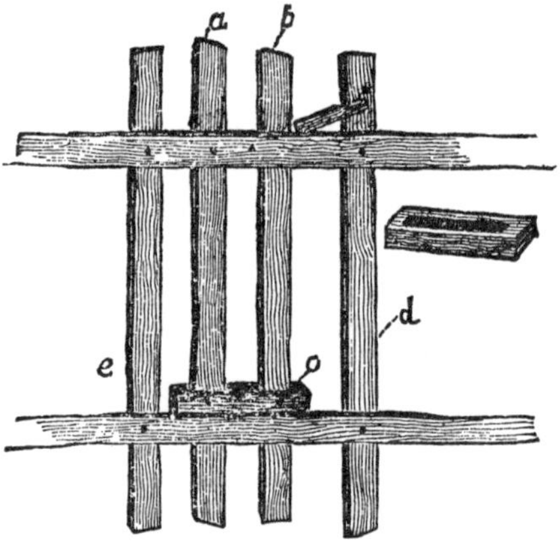

Fig. 11.—STANCHION FOR DAIRY COWS.

should not be screwed up snug, but leave the latch a chance to play. It is usually the plan to set stanchions in a perpendicular position, and if the upper stringer is pitched over against the manger about eight inches, a great gain is made in the ease afforded the animal when it gets up, as its shoulder by this plan does not strike squarely against the latches, and avoids the necessity of "hitching back," to clear the stanchions, and thus prevents the extra strain and exertion often noticed in perfectly rigid, and upright stanchions.

FEEDING CRIB FOR PORK-PRODUCING SECTIONS.

To construct the crib shown in the illustration, four forked posts are set in the ground at the corners of a nine foot square. In the forks are placed stout poles and on these are laid the floor and is built the crib. The posts make the pen high enough for the swine to pass under it; hence, any corn that falls through it is eaten. The feeding floor is laid under and around the pen. In the greatest pork producing sections, nearly all the hogs are fattened from October 1st to January 1st, the corn

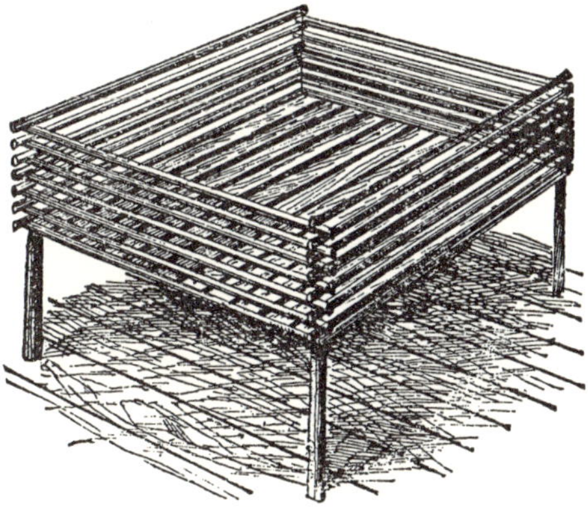

Fig. 12.—CRIB FOR FEEDING LOT.

being fed to the hogs as it is husked. In the pen shown fifty to one hundred bushels can be thrown—enough to feed for two or three days—when it is desired to do other work. It is an easy matter to throw the corn from the crib to the feeding floor, and as the corn will never remain in the crib longer than a week, no roof is required. Set the posts solidly in the ground, for if the weight of the corn should cause the crib to fall, it would kill any fat hogs that might be under it. The hogs cannot possibly get into this crib. Rats cannot infest it. The materials exist on nearly every farm, and any farmer can make this crib and in a short time.

SHEEP RACK AND FEED-BOX.

It is often inconvenient to go among the sheep in feeding them, and there is always trouble from scattering hay or feed about the enclosure or from the animals getting out by the open doors or gates. Figure 13 shows how to feed from outside. The boarding of the pen for about eighteen inches in width, and about six

Fig. 13.—FEED-BOX FOR SHEEP.

inches from the floor is removed, leaving the bottom board in place. Then upright slats are nailed across this aperture inside the fold, allowing twenty to twenty-four inches for each sheep. The slats should be nailed so that an opening eight inches wide is left in the centre of this space for the sheep to thrust their heads through. If much narrower they will rub the wool off their necks.

A tight feed-box with flat bottom and upright sides is made of boards, and placed on the floor outside of and against the slats, and fastened in place. A horizontal swing door, two feet wide and the length of the feed trough, is attached with hinges to the outside upper edge of the feed box. Chains keep it from falling below a proper angle, and a button at the top secures it when closed. The swing door will keep the hay always in reach. With this arrangement one can feed either hay, turnips or grain without going among the sheep, distributing it much more easily than when they are crowding round him. He can also clean out the rack and feed box conveniently from the outside. The sheep cannot crowd each other when eating. When they are through eating, or when the rack is not in use, it may be closed up, shutting off drafts or keeping out dogs. It is desirable to have such an arrangement open under a shed, building or other protected spot, which can generally be provided. It will be found that sheep waste much less fodder and feed than when fed off the ground. The feed trough may be changed so as to come inside the fold, and the rack

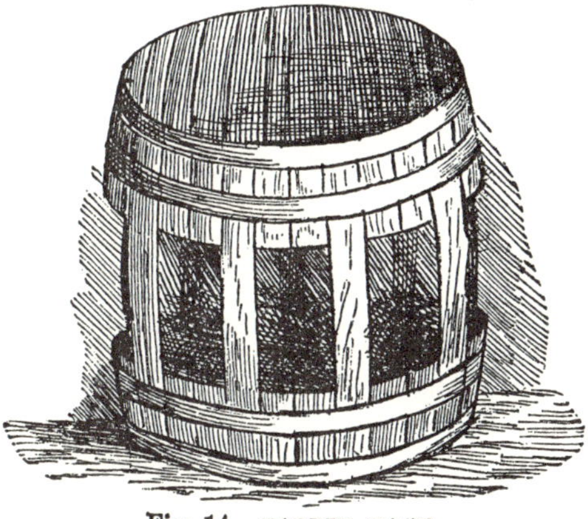

Fig. 14.—BARREL RACK.

made so the sheep can put only their noses through, but it makes the trough inconvenient to reach, and will tend to increase the waste of hay and grain in feeding.

A BARREL RACK.

The illustration, figure 14, shows a rack for feeding hay or straw to calves or sheep. Procure a crockery cask and cut two thirds of the staves, making holes from which the feed can be obtained. If calves are to feed from it, the holes are made slightly larger than for sheep. The animals feeding from this rack waste no food, and the strong cannot so easily drive the weak from it, as from the ordinary rack or manger. Lambs or calves are disposed to fight over their food, and it may be necessary to drive a stake about a foot from the hogshead and opposite the whole staves; this will effectually prevent the weaker ones being driven from their feed. The rack is easily filled, and the fodder, hay or straw may be fed from it without waste; and if moistened bran or meal are mixed with it, forming a complete ration, it may be fed in an economical manner, and be easily reached.

IMPROVEMENTS IN PIG TROUGHS.

One of the simplest troughs is shown in figure 15. The end pieces may be as long on one side as on the other, or

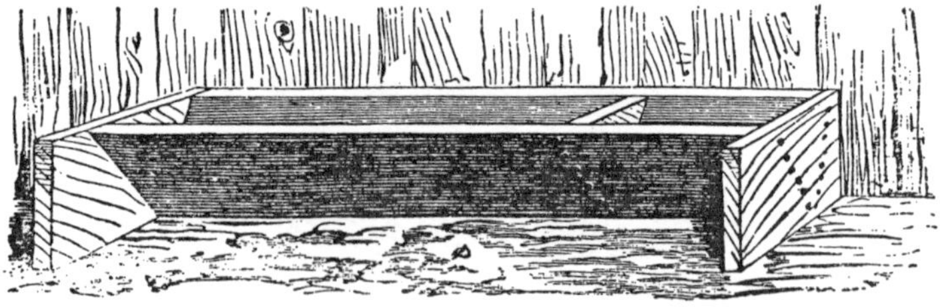

Fig. 15.—SIMPLE PIG TROUGH.

long on one side and shorter on the other, so that the pigs cannot turn the trough over. They may have cross-pieces fastened in strongly every two feet, to makeit less easy for the pigs to stand in the trough, and the trough may stand in the open lot or in positions near the fence.

Where the hogs are confined in pens, a trough is set in the pen as shown in figure 16. This is a fixture, must be strongly made, and be set at the bottom, on a level with the floor of the pen. A pig of one hundred

Fig. 16.—TROUGH INSIDE OF PEN.

pounds weight cannot stand in the trough; the latter can be cleaned out and the feed can be put into it from the

Fig. 17.—GOOD FORM OF PIG TROUGH.

outside. A good form of trough is shown in figure 17.

Here the swinging shutter keeps the pigs away from the trough, or admits them to it, at the will of the attendant. and the trough may be conveniently cleaned out or filled, without any interference by the ravenous herd. Figure 18, shows an improved shutter for the trough last described. The improvement consists of strong bent irons

Fig. 18.—IMPROVED TROUGH WITH SHUTTER.

securely screwed or bolted to the swinging shutter on the inside above the trough, so that a strong pig can neither get into the trough, nor push others away, and get the lion's share. Assuming that ground, soaked or cooked food can only be fed out of troughs with advantage, that pigs will eat and digest well a great deal more cooked food than they will raw, and that the more food a pig eats and digests the more profit there is in feeding him, it is easy to see the importance of good pig troughs.

The engraving, figure 19, represents a good trough for

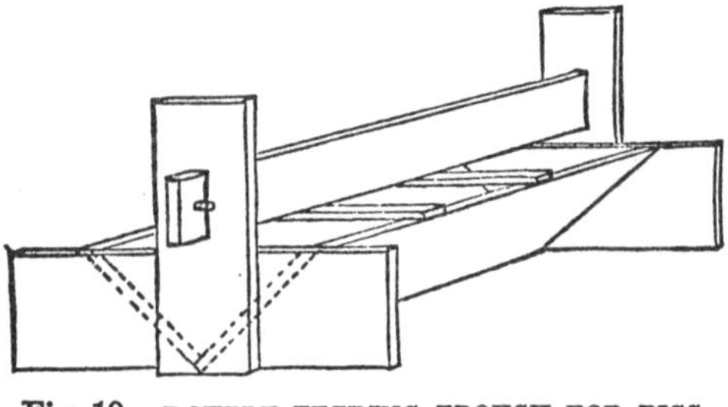

Fig. 19.—DOUBLE FEEDING TROUGH FOR PIGS.

pigs. The sides of the trough are firmly nailed to the end boards. An upright board, which runs lengthwise of the trough, divides it into two parts, and keeps the pigs from getting into the trough. Strips, four inches wide, nailed to the edges of the trough, divide the length into spaces for each pig to feed in, and prevent one pig from crowding the next one. There must always be more spaces provided than there are pigs to feed, in order to avoid fighting among the animals. These troughs may be made of various lengths, according to the number of pigs to be fed.

A PLANK TROUGH.

The common V-shaped trough, as ordinarily constructed, is a short-lived affair. How it may be strengthened and made durable is shown in the engraving, figure 20. The

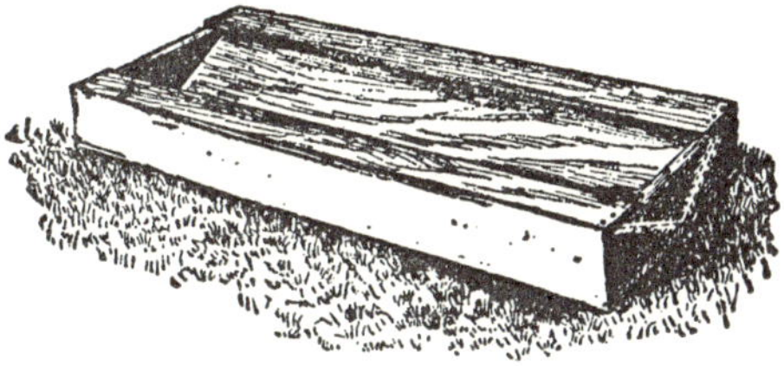

Fig. 20.—DURABLE TROUGH.

trough is made of two-inch pine planks, one six and the other eight inches wide, the end-pieces two inches longer than the extreme width of the trough. Side-pieces of inch pine are nailed at each end, with the upper edge flush and level with the top edge of the ends. A strip of inch pine is nailed from the inside edges of the trough to the outside edges of the end-pieces. When the upper strips become worn, they can be quickly replaced, and there is a hog-trough that will stand very rough usage. The trough should be put together with large wood screws, as these hold better than nails. Place white lead on the joints before fastening the trough together, to prevent leakage. Good tar, applied hot, will answer the same purpose. Some farmers paint the entire trough with hot pitch or tar, which acts as a preservative.

A PROTECTED TROUGH.

Pouring the slop into a trough, with forty hogs crowding and squealing about, is behind the times. When the

Fig. 21.—DEVICE FOR FEEDING HOGS.

slop is thrown into a trough, which passes through the fence to that from which the hogs drink, the stronger ones will crowd together at the conducting trough and get most of the slop. And about every other day a new conducting trough must be made, as the hogs will break it up in crowding for the slop. If it is made to terminate so high that they will not do this, when the slop is poured in, the biggest hog will get directly under it, and the slop, striking on his head and shoulders, will be deflected off to the ground. These evils are avoided by having a separate pen for the trough, filling it, and then letting the hogs in. But it costs something to have an extra pen, and often the space cannot be conveniently made use of.

This device, shown in figure 21, is a rack or screen, made so it will revolve on pins driven through the end-pieces and into the posts, as shown by the dotted line. The trough should be just long enough to fit in between the posts, where it is firmly secured. The most of the trough projects into the hog-yard, leaving merely enough projecting on the other side, to allow of the slop being poured in readily. The illustration represents the frame as it is when the pigs are feeding, and should be hooked into place until they are through. Before pouring in the slop, reverse the rack, so it covers the trough, the extra weight of slats on the hog-yard side keeping it in place until the trough is filled, when the rack is raised and hooked into place, giving the pigs access to their food.

TROUGHS FOR THE PASTURE.

Figure 22, shows a closing trough, nailed against a fence, that is very convenient for feeding bran, oats, corn,

Fig. 22.—FENCE TROUGH.

etc., to cows, calves, sheep and horses. The bottom is made three inches wide, and the outer side stands away from the other, both being set on the bottom. The end pieces of the trough are hinged to the side next to the fence, and the outer side is hinged at the bottom. Strips of leather answer for hinges. A bolt, or strap, passing through the trough at each end allows the outer side to come back just enough to receive the end-pieces, which are held in place by a pin passed through a hole bored vertically through the outer corner of each, and down into the slanting side. To fold the trough up, remove

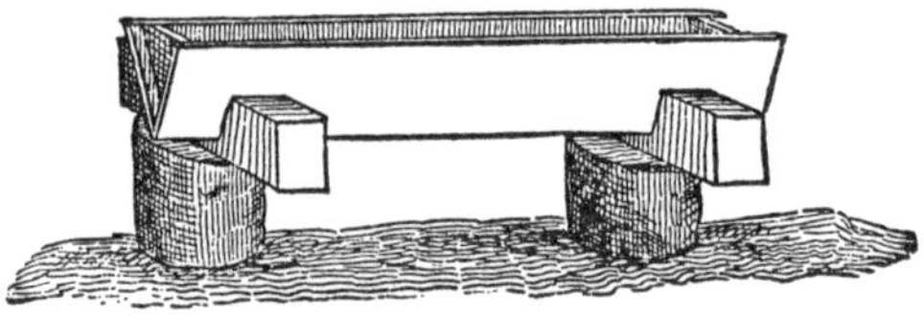

Fig. 23.—A LOW TROUGH.

these pins, and fold the end-pieces inward, bring the outer side up against these, and secure it in place with a strap. This trough is very convenient along the side of a shed, as it can be folded up out of the way. Another closing trough is shown in figure 23. The triangular end-pieces are held in place by cleats on each side. It is

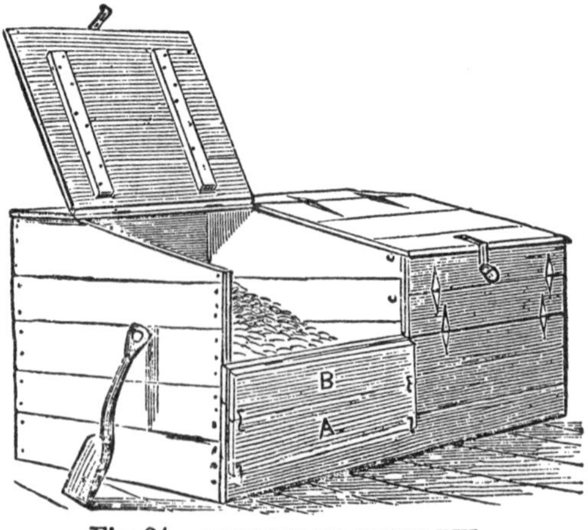

Fig. 24.—CONVENIENT GRAIN BIN.

not necessary to fasten the sides together, but they may be hinged at the bottom. To close the trough, the end-pieces are taken out and laid against one side, while the other side is closed against them. The sides are kept from spreading apart, when the trough is open, by the notches cut in the cross-pieces, upon which the trough rests. These cross-pieces rest upon large blocks.

IMPROVED GRAIN BIN.

A very convenient grain-bin is illustrated in figure 24. The lid or top is raised as usual; then, when desirable, the front top board, which is hinged at the bottom, and hooked inside at the top, is unlocked and let down. This gives convenient access to the bin both in filling and in emptying—enabling one to take out the last remnants of grain or meal.

STRAW BALER.

Good, clean oat straw finds a ready market in cities for filling beds, and other purposes. But its quality and

Fig. 25.—BOX FOR BALING STRAW.

texture are greatly impaired by baling in powerful hay presses, and it is much better, therefore, put up by the aid of a hand press, which preserves the fibre of the straw unimpaired. Figure 25 shows the box and the method of construction. The binding cords are laid cross-wise of the box, resting upon the bottom, as seen in figure 25, and the ends extending through the notches, *B*, *B*, *B*, as shown in figure 26. A small forkful of straw is then

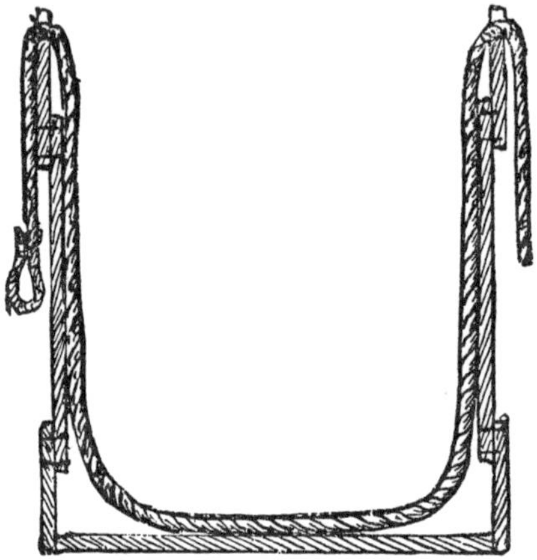

Fig. 26.—MODE OF ARRANGING THE CORD.

placed at each end, and one in the middle, and so on, until the box is filled and the straw packed down compactly. The cords are then brought together around the bundle and securely fastened.

WATERING TROUGHS FOR STOCK.

A good substantial water trough is an absolute necessity on every farm, and we here give illustrations of several useful forms. Figure 27 shows one made of planks or boards. The sides should be of one piece, and also the ends and bottom if possible. If made of two pieces each, joint the edges and join them with dowel pins, using the best white lead between the joints before driving the pieces together snugly. The end pieces should be let into the sides about half an inch, and both the sides and ends should be slightly sloping, so that the form secures freedom from danger of bursting in winter. In putting together, always use white lead on the joints. Use no nails, but draw the parts together with stout iron rods, having large heads on one end and screw threads on the

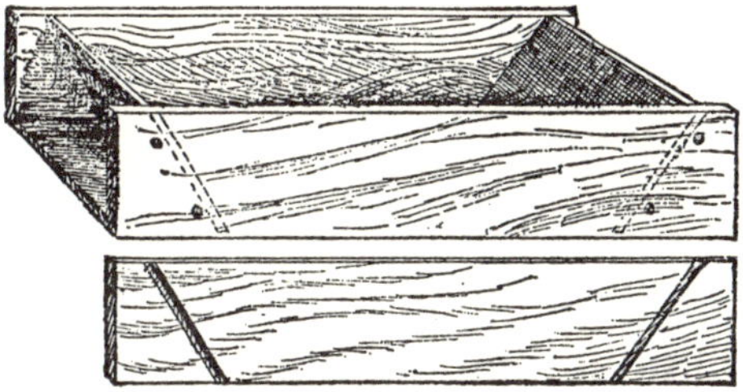

Fig. 27.—A PLANK WATER TROUGH.

other. When this is done, make the bottom edge true, coat well with white lead and securely fasten on with large wood screws. Give the trough a couple of coats of good paint, and when dry, the trough is ready for use. A convenient size is as follows, all inside measurements at the top: six feet long, fourteen to sixteen inches wide and twelve inches deep.

This form of trough will be found useful where water is continually running from water logs, and is designed to prevent freezing and overflow. At one end, as in figure

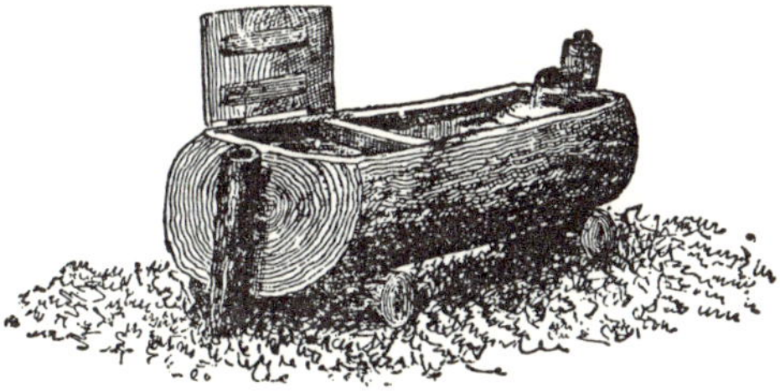

Fig. 28.—WATERING TROUGH.

28, a board is fitted across the trough, and goes to within about one inch of the bottom. The water must flow under this to reach the outlet. This portion of the trough has a cover with a hinge. It will be seen that with this construction no straw or rubbish can get into this covered portion to clog the outlet, and thus cause overflow. This protection is usually sufficient in the winter to prevent the outlet from freezing. But a plug is inserted in the bottom of the trough, which can be taken out when the trough needs cleaning, or in very severe weather.

Farmers who have never used a covered water trough, and who have not been able to keep the water free from leaves and mud in summer, and to prevent the trough

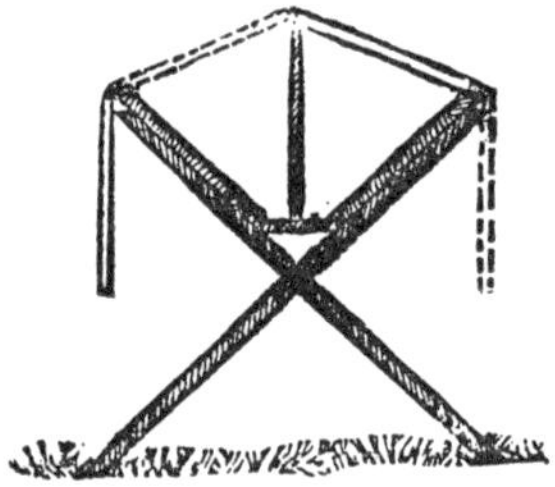

Fig. 29.—COVERED WATER TROUGH.

becoming filled with snow and ice, will be glad of the illustration (figure 29) of a covered trough, which can be used on both sides. It should stand in the middle of a yard, and the best way of supplying it is by a pipe carried underground from a pump. It is supported on crossed posts set in the ground and pinned together. The trough has a central division, upon the top of which the covers rest. When in use, the covers are let down, and when not in use they rest upon the dividing plank, as shown by the dotted lines, and as soon as the stock is watered, the plug is drawn to let the water off.

This non-freezing trough should be made two feet deep, eighteen inches wide, and fourteen feet long, and constructed out of two-inch oak plank. Figure 30 is a sectional side view of the trough. Over it is fitted a double cover, with four-inch space, which extends to within fourteen inches of the outer end. This part is covered with a single hinged cover, which can be raised and fastened up. The trough rests on the ground, and a bank of earth three feet wide is raised around it even

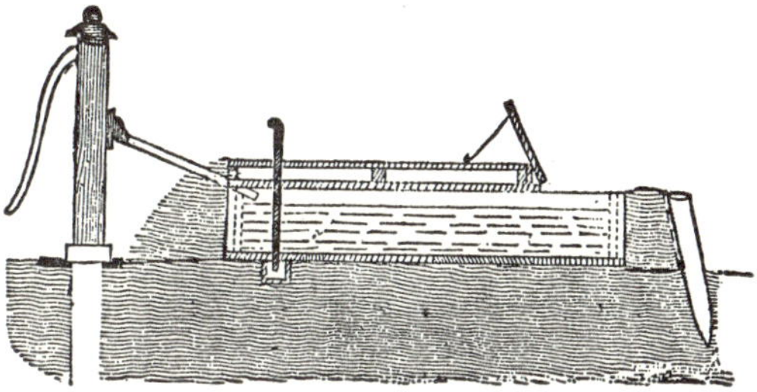

Fig. 30.—SECTIONAL VIEW OF TROUGH.

with the top. At the open end of the trough this bank is eighteen inches thick, and is held up with boards as shown in figure 30. Over all, except the open end, is placed a layer of chaff a foot deep. On the north, west, and south of the well and trough is a tight board fence, one end and side of which are shown in figure 31.

Fig. 31.—THE TROUGH AND SURROUNDINGS.

Across the open end, just back of the opening in the trough, barbed wires are stretched across to keep stock off the well and trough. Under the end of the trough nearest the well is placed a drain, made of fence boards, leading to lower ground. Over this drain is a hole in the bottom of the trough, closed by a plug, which extends through the cover as seen in figure 30, and by which the trough may be emptied into the drain. The trough is filled in the morning, and the natural warmth of so much water having so small a surface exposed, prevents it from freezing during the day, even in the coldest weather. At night the open end is closed. In summer the water in this trough is always cool, and vastly superior for live stock to that standing in open troughs.

A GUARDED HORSE TROUGH.

Chickens have a way of leaving their drinking pens and "fountains," and seeking the more abundant and fresher water of the horse trough. It is all very well so long as

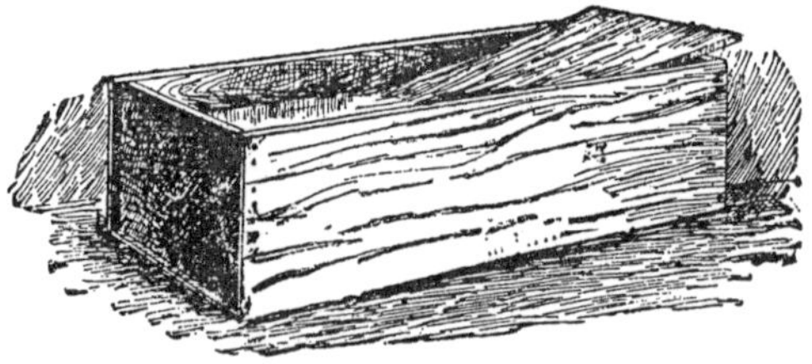

Fig. 32.—FLOATING BOARD IN HORSE TROUGH.

the trough is overflowing, but when the water is low, they lose their balance, fall in and drown. Figure 32 shows a board which floats at one end in the water, and rests at the other upon the end of the trough, being held in place by a twenty-penny nail driven through it. The board, being two inches narrower than the trough, floats freely, and there are no more drowned chickens, for, if they fall in, they can get out again unassisted.

BOX FOR WATERING PAILS.

Figure 33 shows an arrangement for keeping the pails used for watering the horse and cow, assuming that many keep but one or two cows or horses, and that the water is carried to them, from being filled with snow in winter, and from standing in the hot sun in the summer. This plan, as shown in the illustration, is as follows: Have a box standing near the well pump. The size

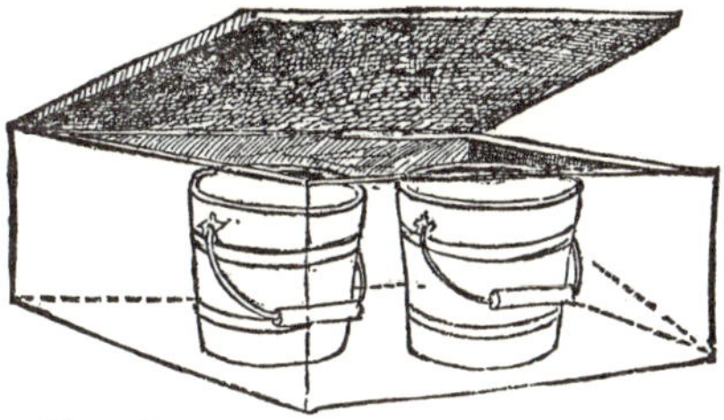

Fig. 33.—BOX FOR WATERING PAILS.

of the box for a single pail should be about sixteen inches square, or twenty inches would be no disadvantage. Have a cover fastened on with either leather or strap hinges; the latter can be bought cheaply at the hardware store, and are better than leather ones. For two pails, the box should be two and one-half or three feet long. In this way, the pails are always in place and much trouble and annoyance is avoided. The best way to arrange the cover is, to have a strip of board some two or three inches in width to go across the top of the box, forming part of the covering, to which the hinges can be securely fastened. Use a smaller box in the hen-house.

HOME-MADE HEATING VAT.

Vats or tanks with wooden sides and metallic bottom, have long been used for heating and evaporating fluids. Figure 34 shows an improved method of construction, which gives greater strength and simplifies the matter of securing water-tight joints. The sides are of pine, two inches thick, ten inches wide, and six feet long. The lower angles are rounded off, as shown in the engraving. Four inches from each end grooves are cut half an inch deep and two inches wide. Into these are fitted and nailed two pieces of pine, two by eight inches, and twenty-five inches long. They are flush with the top, leaving a space of two inches at the bottom. Two rods of half-inch round iron, each with a head at one end and a screw-thread and bolt at the other, are inserted through holes made for the purpose, near the top

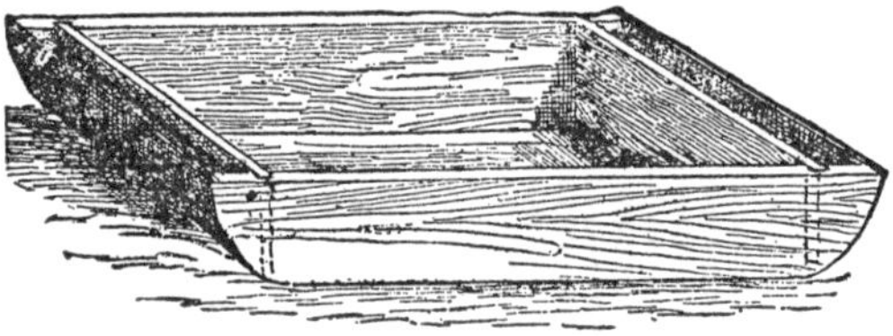

Fig. 34.—VAT FOR HEATING WATER.

of the cross-pieces and screwed firmly in place. The bottom is of galvanized iron, seven feet eight inches long and twenty-eight inches wide. This is fastened by a double row of three-penny nails to the lower edge of the side pieces, extending around the curves to the top. If desired, it may be cut long enough to turn over at the top, and nailed to the upper edge of the wooden cross-pieces. This would give sufficient strength without the iron brace-rods. This vat is set upon an arch of brick or stone two feet wide, so that the wooden sides will project over it. For scalding hogs, a scraping bench is erected close to one side of the vat, and level with the top.

CHAPTER II.

VEHICLES, ROLLERS, HARROWS AND MARKERS.

A CART FOR BREAKING COLTS.

Most colts, if taken young enough, and gently, though firmly handled, can be driven as soon as they can be made to know what is wanted of them. Now and then a spirited fellow feels his oats, or is very nervous about the harness, and still more about the wagon, or cart, and rears, and kicks, and pulls side-ways, trips himself up, and goes down in spite our best efforts to prevent it. For such a good, strong breaking-rig is essential. The cart, figure 35, is home-made, except the wheels; for these a pair of strong wagon wheels—either front or hind—will do. The shafts are a pair of seasoned hickory poles, extending about two feet behind the wheels. They are bolted upon the axle-tree, and underneath these is a lighter pair of poles, attached to the shaft in front, and bolted also to the axle-tree by the same clamps that are used to hold the shafts. These extend back as braces. and are mortised into pieces, which are themselves mortised into the shaft-poles near the end. The object of this arrangement is to keep a colt from rearing. The ends of these pieces will bear upon the ground, the moment he lifts himself up. The same result would be accomplished by having the poles extend far out behind, but this makes turning exceedingly awkward, so that rigs of this kind can only be comfortably used in an open lot. The box, or body of the vehicle, is made with reference to strength, so that it cannot easily be kicked to pieces, nor broken by overturning or being run away with. A strong plank is bolted to the poles in front; uprights, and cross-boards of three-quarter-inch spruce, form the dash-board, which is well braced. The back and seat are similarly attached. It is important that the seat should

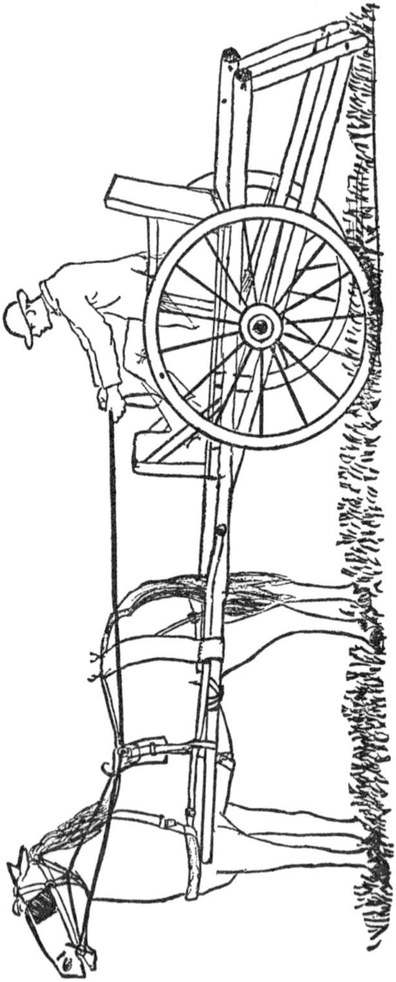

Fig. 35.—A RIG FOR BREAKING COLTS.

be so placed that the driver may at will throw his full weight forward, to bring the bearing of the shafts upon the saddle, or backward, to lift up on the girth or belly-band. The harness should always be sufficiently strong, and before using the breaking-cart, the colt must be well harness-broken.

A HOME-MADE CART.

Figure 36 shows a serviceable farm cart, which can be made by any one who understands the use of a saw and hammer. The sides of the box, which is six feet long and four feet wide, are of plank a foot wide, the bottom of inch boards; the end-board is fastened with hooks, so that it can be readily removed when loading the cart.

Fig. 36.—HOME-MADE CART.

The wheels are those of an old, worn-out reaper, and the axle consists of a piece of gas-pipe, large enough to fit the hub of the wheels. Pins put in holes drilled through the ends of the axle, keep the wheels in their places. The axle is fastened to the wagon by wooden blocks, hollowed out to proper shape; these blocks are firmly screwed to the side-pieces. The thills pass through the front board and are bolted to the sides of the box. A single-tree is fastened to a cross-piece bolted to the thills close to the box. Such a cart is very convenient on every farm, and being low, it is easily loaded.

APPARATUS FOR LIFTING A WAGON-BODY.

To lift a heavy wagon-body from its truck is tedious work, if to be done by main force only. The use of pulleys facilitates the operation materially, but not as much as the apparatus shown in figure 37. It is simple, very convenient, and may be easily made by any farmer handy with tools. *b*, in the engraving, is a wooden roller, about three inches in diameter, and resting on the

Fig. 37.—APPARATUS FOR HOISTING A WAGON-BODY.

joists *a*, which are over the wagon in its shed. *d* is a rope which winds around the roller, and is fastened at its lower end to the cross-piece *e*. Through each end of the cross-piece passes a half-inch round iron bar, *f*, with bar on top of *e*. The lower ends terminate with square bends of three inches, which hook under the box, and when turned half round will slip off, and may be hoisted up and put out of the way. The handles, *c*, are four feet long and are mortised into the roller. A man or boy standing on the ground can turn the handles with ease, and raise the box from its bed in half the time four men could do it by hand.

JACK FOR WAGON BOX.

A cheap method of removing a wagon box is shown in figure 38. A platform to receive the box is made by driving stout stakes into the ground and nailing cross-pieces to them. The platform should be as high as the top of the wagon standards. The lifter consists of a stout piece of timber, which will reach two feet above the wagon box, the top rounded, and a pin, driven into it, which passes through a slot in the lever. Two chains, provided with hooks, are fastened at the short end of the lever, and a rope at the other. One arm of the lever is

Fig. 38.—JACK FOR WAGON BOX.

three feet long, and the other nine feet. The wagon is driven close against the side of the platform. The lifter is placed, as shown in the engraving, on a line midway between the wagon and the platform. The hooks on the end of the chains are caught under the box, or the rod which passes through the rear end of the box, and by pulling on the rope, the box is easily lifted out and swung around on the platform. Then lift the front end over. The jack can be used to return the box to the wagon. The pieces need not be large, and when made of seasoned wood, the jack is easily handled.

SERVICEABLE WAGON-JACKS.

Take a scantling two and a half feet long, one inch thick, two and a half inches wide; rip it with a saw from top, to within five or six inches of the bottom, like a

Fig. 39.—WAGON-JACK.

tuning fork, figure 39. One prong is the lever, saw the other prong off at top, one inch higher than the bottom of the hind axle; then saw it off at the shoulder five or six inches from bottom; fasten it on again with a hinge exactly where it was sawed off, and it is ready for use.

Fig. 40.—HOME-MADE WAGON-JACK.

Set it under the axle, lowering the lever enough to allow it to go there; then raise the lever past the balance, and it will go together of its own weight, and stay there. At the left of the engraving it is seen as lowered, at the right as raised. This jack is very cheaply made, and varies in dimensions according to the weights to be raised. In the one shown in figure 40, the lever *a* is made of one-inch stuff, and the post *b* and the bearing-piece *c* of two and a half by two and a half. The latter two are slotted to admit of the lever working freely in them. The bearing-piece is held to the lever with an iron or a wooden pin, a little behind the post or fulcrum, so that when in use the jack will support the wagon without any other fastening.

ADJUSTABLE WAGON SEAT.

A six-inch board has slots cut in each end, so as to go between the stakes of the wagon. Another board, one foot wide and three feet long, is fastened to the first in the position shown in the engraving, figure 41. An old seat, from a harvester or mower, is fastened upon the boards, when an easy and satisfactory seat is provided for a wagon when in use for purposes of drawing wood, lumber, etc.

Fig. 41.—A WAGON SEAT.

LUBRICATING AXLES.

Many lubricate axles only to prevent wear; they overlook the fact that by reducing the friction they lessen the draft. A well-oiled axle lightens the load. Oil to axles is best governed by the rule of "little and often." If too much is used it exudes at the ends, gathers dust, and thus the lessening of the friction is not so great, while oil is wasted. In nearly every case where the lubricant is wasted it is because it is stuff not fit to be used, for a good lubricator costs enough to keep the average man from allowing it to waste. Oil that "gums" much is unfit to be used. Castor oil is a splendid lubricator for axles, but used alone may gum too much. This is corrected by the addition of refined coal-oil (that used for lamps), or lard; the coal-oil is the better. Some wagons are yet made unprovided with metal shields or "thimbles," being banded with steel; for these some tallow may be used, as it is one of the best of lubricants when iron and wood are brought together. Pine-tar is a good addition to the lubricant for wagon axles and is a part of most of the "axle greases" sold. Plumbago is another good addition; its fine particles fill the small irregularities in the opposing surfaces, thus making them smoother. A mixture of lard and plumbago is good for the journals of reapers, mowers, etc.; we have found castor oil and refined coal oil also good for this use, particularly for use on the "sickle-driver." For carriages nothing is better than castor oil and a very little lard oil or refined coal oil. Lard oil alone has not "body" enough for the journals of reapers, mowers, etc.; add a little castor oil, or tallow or plumbago. While the axles of reapers, grain-drills, hay-rakes, etc., will not need lubricating so often during the year as the axles of the wagon, oiling them must not be neglected, as the rough ground the wheels pass over makes the wear on unoiled axles quite rapid. The axles of corn-cultivators require frequent lubricating. For these the best lubricants are those recommended for wagon axles.

A LIGHT SLEIGH OR "JUMPER."

A light sleigh may be made of hard-wood poles cut and bent into shape, a few bolts, and a light body or box. Figures 42 and 43, made from sketches of a recently constructed "jumper," will serve as a guide to any one who wishes to provide himself a light sleigh at a trifling cost. Two hickory poles, for the runners, are dressed down, and the small ends bent to the proper curve and fastened until they will retain the bent shape. The posts are mortised into these runners and the bench pieces, which latter are firmly fastened together with bolts. The braces and their positions are shown in the engravings.

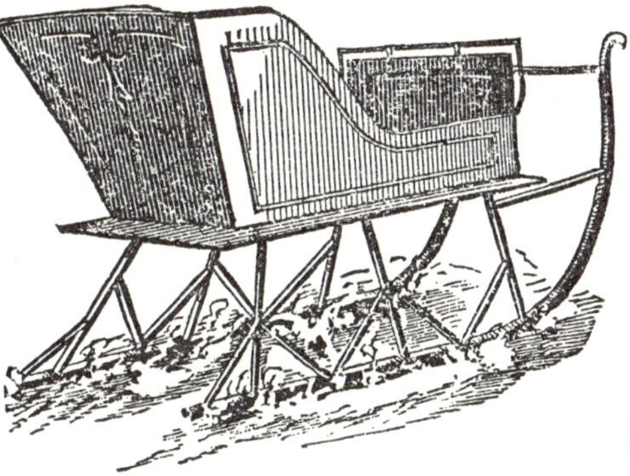

Fig. 42.—A JUMPER.

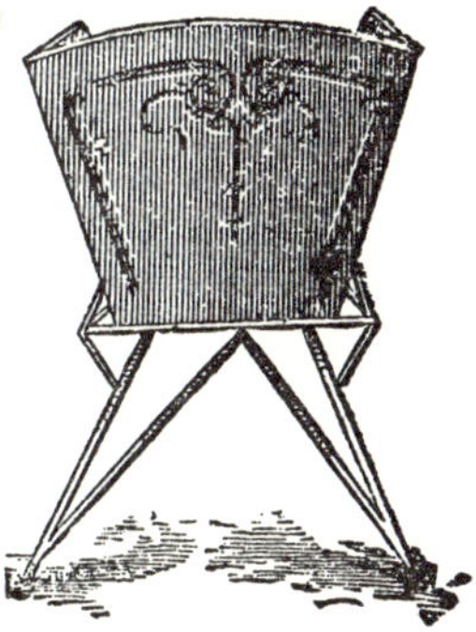

Fig. 43.—REAR VIEW OF JUMPER.

A floor is laid upon the bench pieces, and extends beyond the sides of the box or body. The box may be plain or ornamented in various ways. The one shown in the engraving has the sides and back flaring. The shafts are fastened to the curved end of the runners with eye-bolts.

A SUBSTANTIAL SLED.

Figure 44 shows a sled which is principally used in the pineries of Michigan, where a single team will draw on it from two to five thousand feet of lumber in the log. Special roads are kept open to accommodate these broad-track sleds, and when a load of a dozen or more logs is under way, it would be perilous for any who should venture to block the road.

Figure 44 shows the general construction of the sled. The bunks, *a, a,* are eight by ten inches and ten feet in length; the sway bars *b, b,* are four by four inches; the reach, *c,* is ten feet between the bunks, the beams, *d, d,*

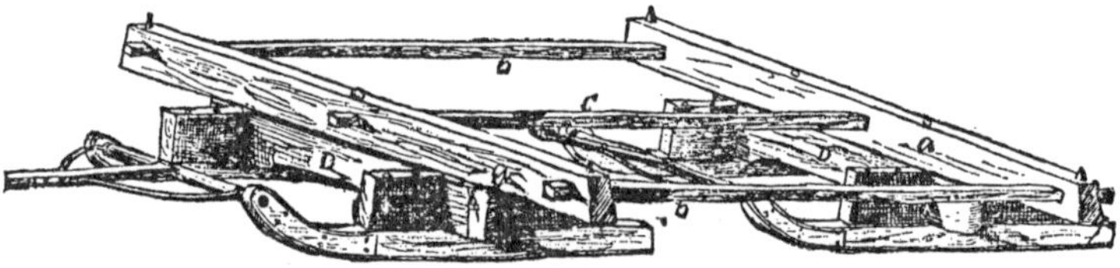

Fig. 44.—MICHIGAN SLED.

are ten by twelve inches, and the track is four feet eight inches long.

The particular feature of this sled is the concaves, *x,x,* made in the beam, *F,* which fit two convexes in the block, *E,* as shown in figure 45. These taper from the top to the bottom, fitting snugly at the bottom, and open one-sixth of an inch on each side at the top. By this

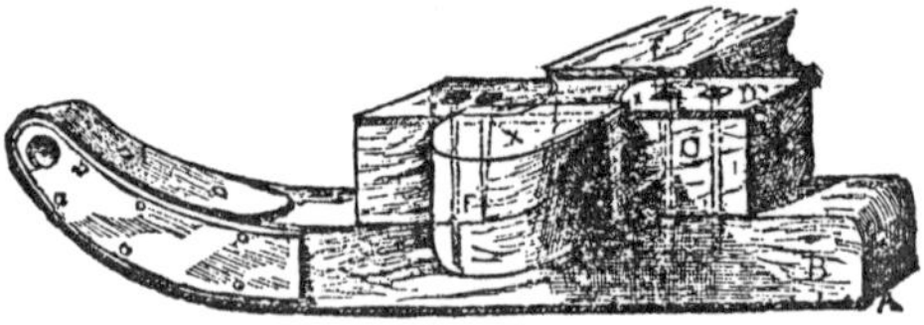

Fig. 45.—SLED RUNNER.

means slight play is allowed to the runners, which eases the motion considerably on rough ground. *A,* in figure 45, shows one of the steel shoes which are four by five-eighth inches; the runners, *B,* are four by six inches, and four feet long; the blocks, *C,* are four by twelve inches, and three feet in length. The iron plates are shown at *D,* the bolts at *C;* the beam, which is ten by twelve inches, at *F.*

A DUMP-SLED.

A method of constructing a dump-sled for hauling manure, earth and other substances, is shown in figure 46, and it will be appreciated by many northern

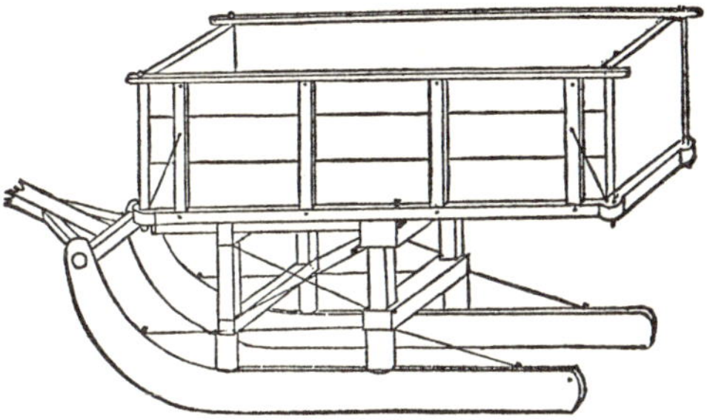

Fig. 46.—A SLED ARRANGED FOR DUMPING.

farmers. The front bob of a double sled has the framework raised by means of a trestle, and upon this the box is secured by eye-bolts, and a staple and pin.

A TRIPLE LAND ROLLER.

A great objection to the use of the roller is, that it tears up the ground for a considerable space when it is turned around. Another is, that the weight of the tongue and frame bears heavily upon the necks of the horses, and often causes sores. The roller shown in figure 47 has neither of these objections. It is made in three sections, and the hinder section balances the weight of the frame and tongue. In turning, the whole implement moves easily with the side roller as a pivot, and avoids all disturbance of the soil. The center roller is made a little longer than the side ones, and thus secures the complete pulverization of the soil. The rollers are easily made, either of solid logs, or of round discs, to which narrow bars are spiked. The best roller is the heaviest, and cast iron is the best material; although much cheaper ones may be made of artificial stone molded in wooden cylinders. The material may be mixed as follows: One barrel of good hydraulic cement is well mixed dry with three barrels of coarse, sharp sand. A sufficient quantity of the mixed cement and sand for one section is then wetted and worked up into a thin mortar, and is at once put into the mold; broken stone, first wetted, may

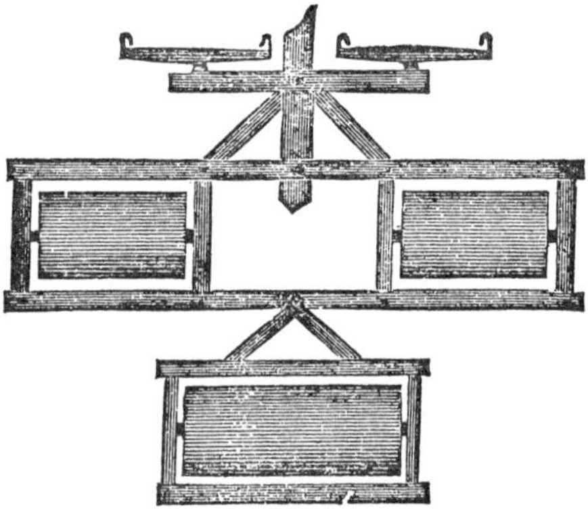

Fig. 47.—A TRIPLE LAND ROLLER.

be worked into the center, around a square shaft of oak timber, carefully centered. The whole is well rammed down, and more is added and rammed as it is put in, until the mold is filled. The ends of the roller should be of clear cement and sand for a few inches, only the interior being filled in with stone for the sake of economy and for weight as well. When the mass is dry and solid, the mold is taken apart. Wing gudgeons are fitted into the oak shaft. They run in wooden boxes, bolted to the under side of the frame. In this way a most excellent and useful roller, equal to a cast iron one and quite as durable, may be made for a cash outlay of about three dollars only.

A CHEAPER TRIPLE ROLLER.

Figure 48 shows a much simpler form of triple farm roller, made chiefly of wood. It is in three sections, each about two feet long, such a one being much easier on the team than when made solid or in merely two sections. A good oak or maple log, as nearly cylindrical as possible for ten or twelve feet, can be cut in the woods, the bark peeled off, and the log sunk under water for several weeks, when it is to be dried out under cover.

Fig. 48.—FARM ROLLER.

If seasoned with the bark on, the worms are apt to work on it. Saw off the pieces the required length, strike a center and work them to a uniform size, and then bore holes for the journals. The best way is to have a pump-maker bore entirely through the pieces an inch and three-quarter hole. Then hang them on a round bar of iron or steel, an inch and a half in diameter, as a loose spindle. The brace-irons can be made of stout old tire by the nearest blacksmith, and four of them, securely bolted into place, will be sufficient. Keep under cover when not in use.

A DOUBLE LAND ROLLER.

The cheap home-made roller shown in figure 49 consists of two sections of a round log, dressed smooth, and fitted in a frame. The frame is made of four by four oak, bolted together firmly. The logs are each eighteen inches in diameter, and three and one-half feet long, one being set three inches ahead of the other in the frame. The pins for the rollers are one and a quarter inch thick, round for four inches at one end, and square

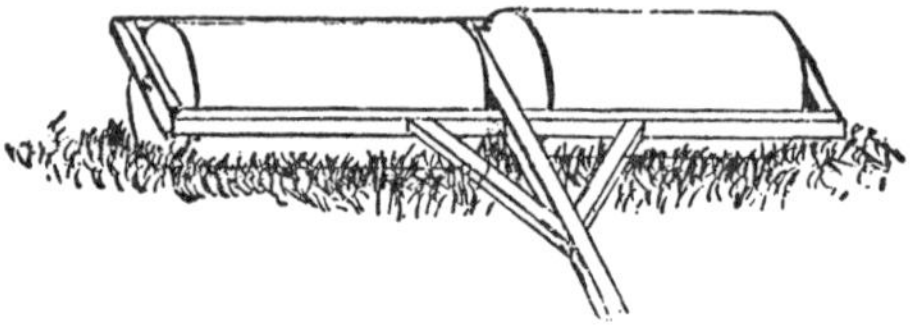

Fig. 49.—A HOME-MADE ROLLER.

for twelve inches; this end is pointed, and is driven into an inch hole, bored in the end of the log. The tongue is braced with strong iron braces, and a seat may be fitted partly over the rear of the frame, and balance the weight of the tongue, and relieve the horses' necks.

STALK LEVELER.

The frame, figure 50, is of two pieces six inches wide and two inches thick. They are joined together with

Fig. 50.—STALK LEVELER.

pieces of old wagon tire, which has been straightened out, and two holes punched or drilled in each end, to hold the spikes. The front ends of this tire-iron are bent or curved, to hold the chain to which the horses are attached. By using this contrivance when the stalks are stiff and hard with frost, they will break off clear and clean near to the ground, and can then be gathered up and burned, or made into manure.

USEFUL CLOD CRUSHER.

The illustrations, figures 51 to 53, present different views of a home-made implement to be used as a clod crusher

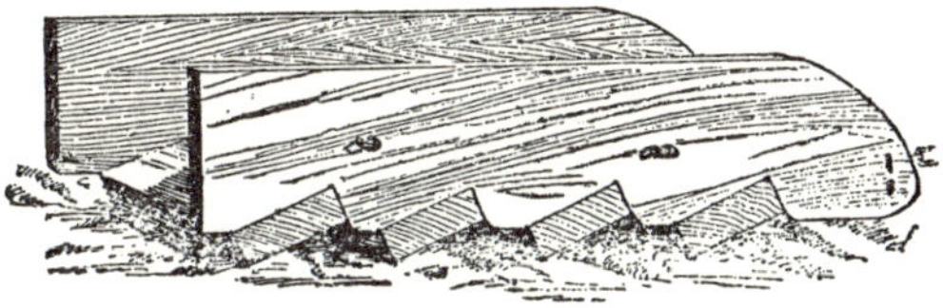

Fig. 51.—CLOD CRUSHER IN OPERATION.

or for other purposes. The runners are of oak plank, two inches thick, six feet long and eight inches wide, each rounded off at one end, and notched on the upper edge, as shown in the engravings. The cross-pieces are of similar material, three feet long and seven inches wide, spiked in place. The outer edges of the cross-

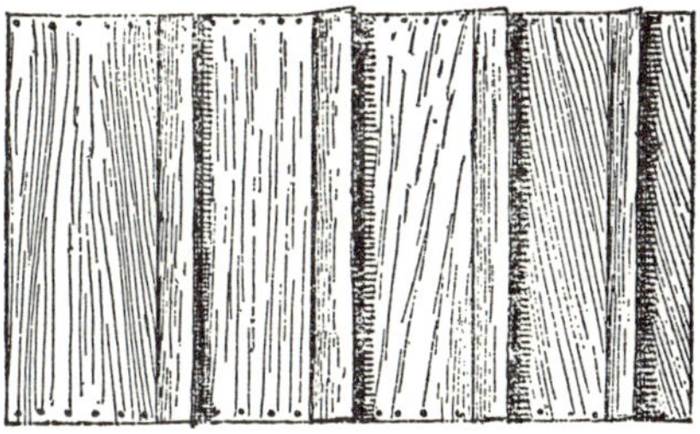

Fig. 52.—BOTTOM OF CLOD CRUSHER.

pieces are faced with band-iron. A staple with ring is driven from the inside of each runner, near the front, and the chain by which it is drawn is run through the ring. In this form it serves a very good purpose as a clod crusher. If additional weight is desired, large stones may be placed between the runners.

To fit it for use as a sled, it is inverted, a box of inch boards made five feet ten inches long, three feet broad, and nine inches deep. The lower edges of the side-boards are notched to fit the projections of the cross-pieces. Inch boards are nailed across the bottom to close the spaces between the latter. Staples are driven into

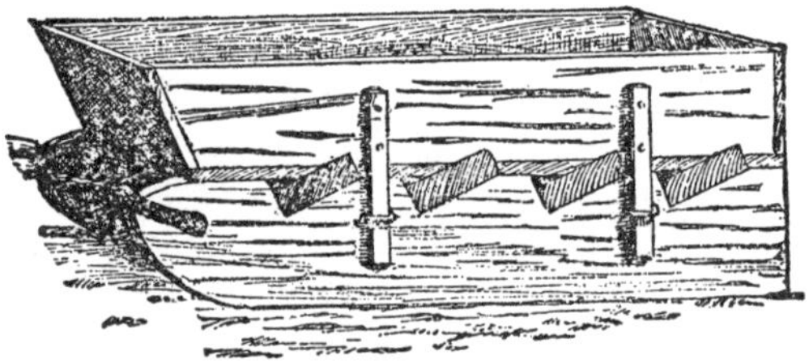

Fig. 53.—CLOD CRUSHER AND SLED.

the sides of the runners to receive hickory stakes, which hold the box in place. For use in winter the thills are attached by iron straps bolted on, as shown in figure 53. When the runners become worn, the bottoms are planed off and strips of oak pinned on. The box may be replaced by a rack for drawing hay or other bulky stuff.

A BRUSH HARROW.

For the cultivation of various kinds of crops, one of the most useful implements made on the farm, and one which properly constructed, lasts a lifetime, is a smoothing and brush harrow, figure 54. It should be made of rather heavy stuff, so that the weight, as it is dragged along, will be sufficient to break the lumps and level the soil. This harrow can be used with good effect in covering newly planted seed, and in all cases where a disc or tooth harrow would be too heavy or wide-spread, a brush

Fig. 54.—BRUSH HARROW.

harrow, like that herewith represented, will be found to be a good substitute.

AN IMPROVED HARROW FRAME.

Figure 55 shows a very cheap and excellent harrow frame intended for grass seeding; also for working

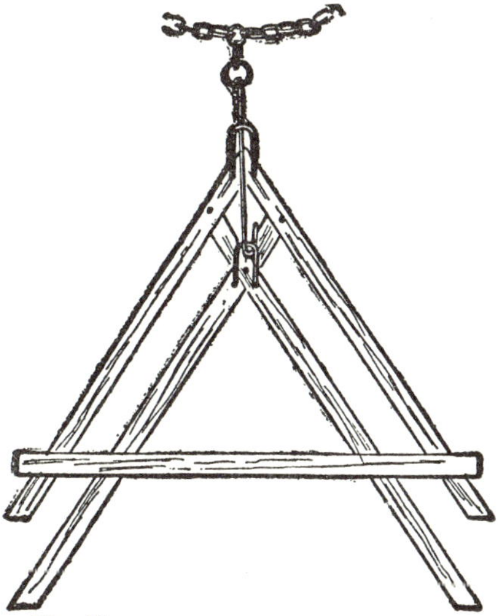

Fig. 55.—IMPROVED HARROW FRAME.

corn and potato land while the crop is young and small. For this purpose, a harrow should be light, broad, have a large number of fine teeth sloping backward, and should be so arranged that it will draw level and not lift at the front. The owner and inventor of this harrow claims that he has secured all these. The special point of this harrow is the hitching device. This consists of a hooked bar which works in two stirrups, one to draw by and the other to permit the draw-bar or chain to rise and fall, as the harrow passes over the ground that is not quite level. This is an important end to secure. The harrow is not patented, and any farmer is free to make one.

LAND-MARKERS.

Figure 56 represents a one-horse land-marker, such as is used among the gravel and cobblestone soils of some

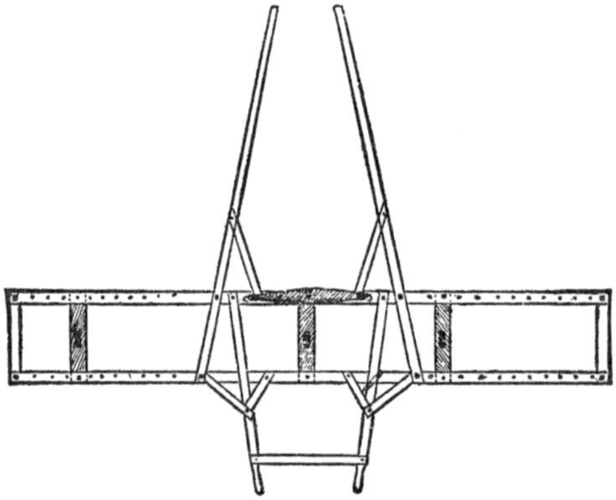

Fig. 56.—LAND-MARKER COMPLETE.

sections, where it does good service. The lumber should be of well seasoned oak; the long rails, two by three stuff in pairs; the cross-bar and end pieces the same; the cross bars, in which the teeth are set, three by three inches square; the thills one and a half by two inches at the large ends and tapering beyond the braces. The handles are common straight plow handles, that is, bent only at the grip. Three-eighths bolts are large enough for the frame.

The center tooth should be permanently framed in, the outside teeth being adjustable, work in the slot between

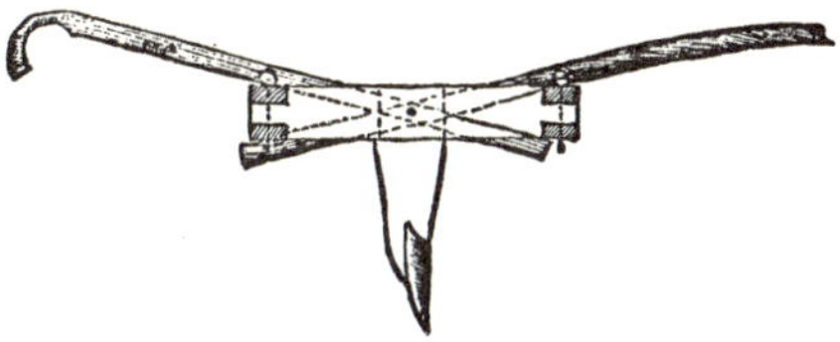

Fig. 57.—END VIEW OF LAND-MARKER.

the long rails, and are held in place by two three-eighths iron pins. They can be moved so as to mark from two feet six to five feet. The rails should have seven-sixteenth holes bored through them every three inches, commencing at two feet six from center of middle tooth. For shares use old points of shovel plows. The whiffletree is held by a bolt which passes through the center cross-bar.

Figure 56 shows the adjustment of the teeth, one being set at two feet six, the other four feet, also the position of the thills, the whiffletree, the handles. The cross-rail tenons at ends should fit in the end of slots and be bolted

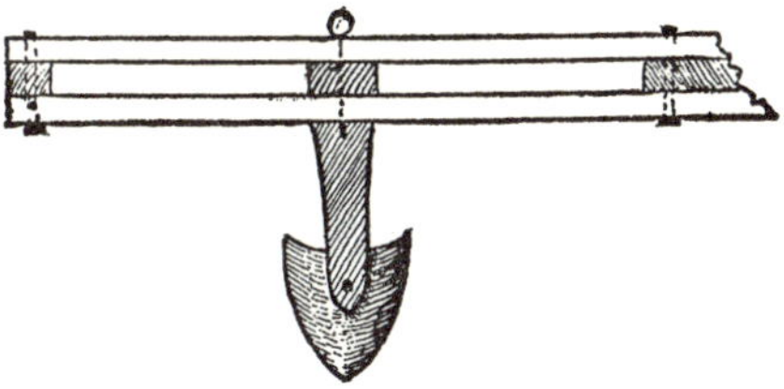

Fig. 58.—MOVABLE TOOTH OF LAND-MARKER.

fast with three-eighths bolts. The braces on thills and handles are of iron, a quarter of an inch thick and an inch wide, held by quarter-inch bolts. Figure 57 is an end view, showing the pitch of handles and thills, a tooth also, and the mode of fastening the same. Figure 58 shows one end of the pair of long rails which form the slot for a movable tooth; also the shape of share. This

Fig. 59.—A HORSE LAND-MARKER.

implement is not patented, and can be made by any one with common tools and knowledge and ingenuity enough to use them.

Figure 59 shows a marker with plank runners, so simple in its construction and so clearly shown in the engraving that no description is needed.

Figure 60 shows an excellent marker for "checking" corn ground. The runners are of hard-wood plank two by six inches, and four feet long. They are usually placed three feet ten inches apart. The cross-pieces, of two by four inch stuff, are laid on top of the runners, and fastened in place with square pieces; or better yet, let into the runners. Pieces of two by four inch stuff run diagonally from the rear corners and meet in front, forming bases of attachment for the pole tongue. Bows of pieces of hoop-poles are fastened in these, through which the rear end of the tongue passes. This is much superior to bolting the tongue across the top of the marker, for then every irregularity in the walk of the horses is communicated to the marker, making short crooks in the checks; and where the marker dips in a depression, its weight is

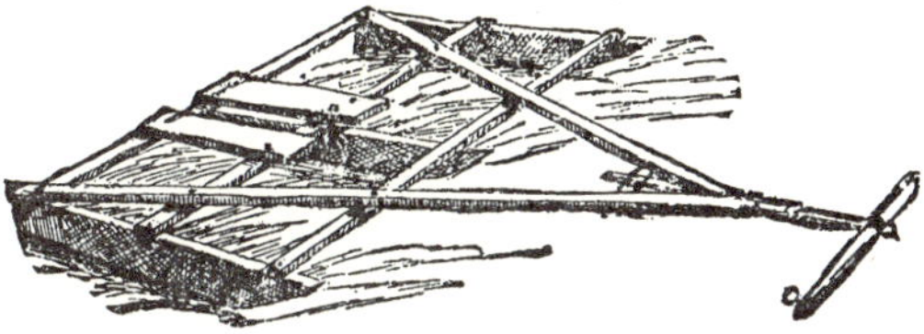

Fig. 60.—SERVICEABLE MARKER.

thrown on the horses' necks. When the tongue is attached, as shown in the cut, no short crooks are made in the checks, there is neither lateral or horizontal strain on the horses' shoulders, while the hoops make the marker manageable in crossing deep furrows, etc. The tongue is held in place by a round iron bolt passing through it and the end of the diagonals. The double trees are fastened just in front of this point of attachment. The driver stands on the two boards on the rear center of the marker.

COMBINED MARKER AND CLOD-CRUSHER.

In figure 61 is a very clear illustration of a useful marker and clod-crusher, which is made as follows:

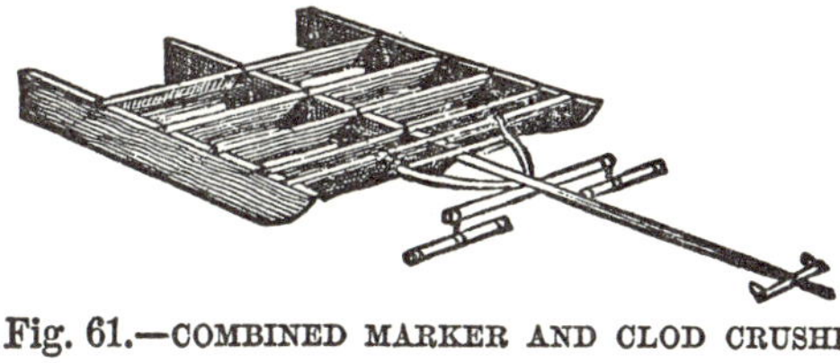

Fig. 61.—COMBINED MARKER AND CLOD CRUSHER.

Three runners are provided, four feet long, eight inches wide and two inches thick; four two-inch planks of strong, hard wood, eight feet long and eight inches wide, are let into the runners four and one-half inches deep; these slope from top to bottom edge backwards, forty-five degrees, so as to draw over the rough ground, and break clods by pressing on them. These runners are let into the cross-pieces one inch, and are fastened together by large screws. A strip of two by four is halved down on the runners in the front, for a draw-bar. The tongue is fitted with hooks, which are attached to rings on the draw-bar, so that it can be removed when the sled is turned over to be used as a clod-crusher.

A LAND LEVELER.

For preparing land for grass seeding, or for corn-planting, the three plank leveler and clod-crusher is useful.

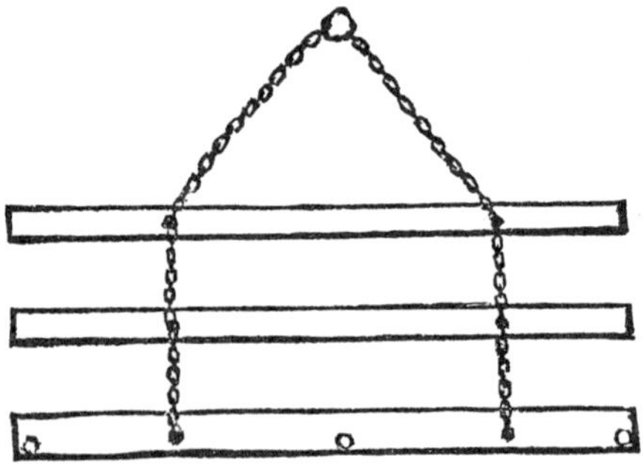

Fig. 62.—LAND LEVELER.

The planks are held together by a chain, and both with large washers, which pass through links. If short pieces of heavy chain are fastened to the holes in the rear plank they will mark sufficiently plain for corn or potato planting.

CHAPTER III.

SMALL TOOLS AND APPLIANCES.

BAG HOLDERS.

There is an endless variety of devices for holding a bag upright, with the mouth open. One of the simplest, figure 63, consists of a piece of hickory or white oak bent into a half-circle, and the ends passed through a

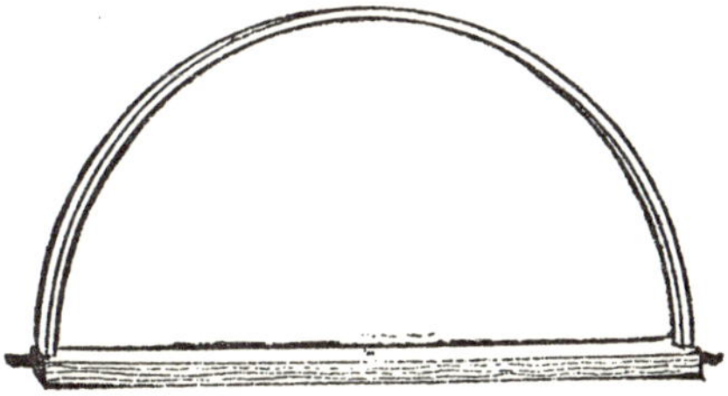

Fig. 63.—CHEAP BAG HOLDER.

somewhat larger rod of the same kind of wood, and wedged fast. A screw is driven into each end of the rod, and filed to a point. To use it, the mouth of the bag is put through the half circle, and the edge is turned down over the holder, and over the sharp points, which hold it firmly. The bag is then held while it is filled, or it may be hung upon two hooks, or the holder may be fitted in a frame on a stand, so that one can use it without any help to hold the bags.

A very good form is shown in figure 64 for farmers who sack their grain in the granary, one side of the room being used as a passage-way. It is swung by staples to the posts, and can be changed readily from one post to another by having staples arranged in each post. Three-quarter inch round iron is used, all in one piece, the rod being bent or welded to make the circular shaped opening for the hopper. The hopper is made of common sheet iron, funnel-shaped, turned and wired on the upper

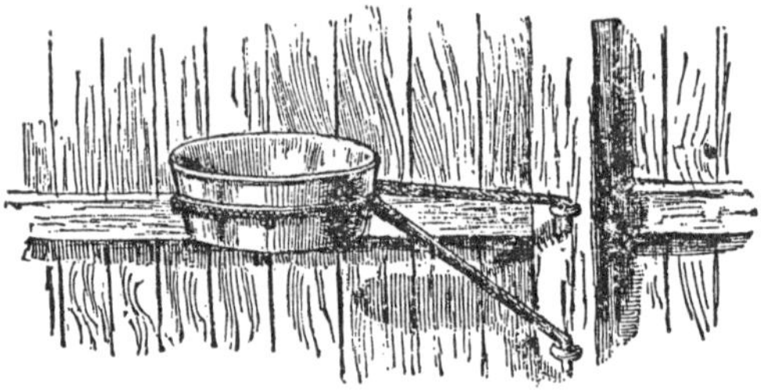

Fig. 64.—GRANARY BAG HOLDER.

side to add to its strength and to reduce the sharpness of the edge. Four small hooks can be riveted to the hopper, to attach the sack when filling it. When not in use, the holder can be swung back out of the way. If desired, the hopper can be permanently attached to the iron rim or holder by a couple of small rivets passing

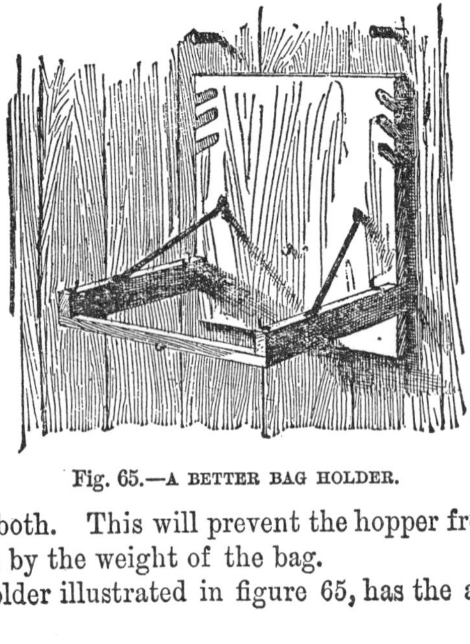

Fig. 65.—A BETTER BAG HOLDER.

through both. This will prevent the hopper from being displaced by the weight of the bag.

The holder illustrated in figure 65, has the advantage

of being built almost wholly of wood, and can be made by any ingenious farmer. It can also be adjusted to various heights by moving it up or down a notch. The back is of inch board, about one foot wide and of any desired length, from fifteen to thirty inches. The arms are an inch thick and an inch and a half wide, fastened by screws into the notches in the back and supported by wire rods which may be held by screws through the flattened ends, or may pass through the back and arms and clinch. The cross-piece is of tough wood, three-fourths

Fig. 66.—PORTABLE SACK HOLDER.

of an inch square. For holding the bag there is one hook on the back piece, two on each arm, and one under the cross-piece. The whole is supported on two strong spikes driven into the wall of the barn or other building, and projecting far enough to fit the notches on the side.

The bag holder shown at figure 66, is portable and may be taken wherever it is to be used. The sack to be filled is brought up inside of the frame and turned over and hooked on the underside of it. The hooks are put here because they are not in the way and the sack is not torn by the weight of the grain, as would be the case if the hooks were put on the top of the frame. The frame must be somewhat smaller than the sack. The sack can be filled to the top of the frame, as the part drawn over will be enough to tie by. The material used is inch stuff. The length of the legs must be such that when the sack is put on the hooks the bottom will rest on the floor.

Another form of portable holder, shown in figure 67, is so compact and light that it can be carried into the

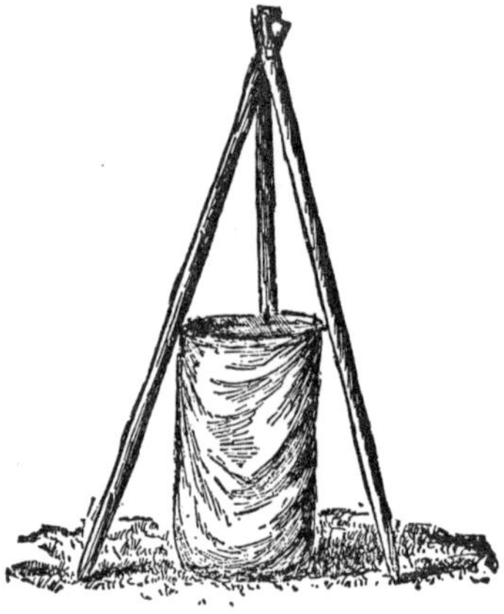

Fig. 67.—A SIMPLE BAG HOLDER.

field if desired. The apparatus consists simply of three light poles about six feet long, and loosely fastened together at one end with a small carriage bolt, and three screw-hooks at the proper height for holding the bag when stretched out, as seen in the illustration.

HANDLING POTATOES.

Potatoes are best stored in a dry, cool cellar, where the temperature can be kept by ventilation at about forty degrees. The floor should be of planks, raised three inches from the ground, and laid with one-inch spaces between them for ventilation. The bins should be about eight feet long, four feet wide and deep, made of loose-barred partitions (figure 68), wired together at the corners. A bin of this size will hold one hundred bushels, and with such a one it is very easy to know precisely how much the crop amounts to.

The box shown at figure 69 will be found a great convenience in gathering and storing the potatoes. It is made eighteen inches long, fifteen inches deep in the clear at the sides, and ten inches wide, all inside measurements; thus holding two thousand and seven hundred

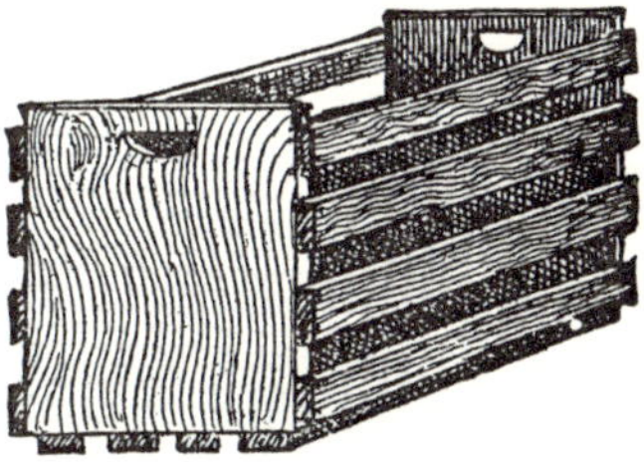

Fig. 68.—PANEL OF POTATO BIN. Fig. 69.—SLATTED BOX FOR POTATOES.

cubic inches, or thirteen cubic inches (about two good-sized potatoes) over a heaped bushel, which is two thousand and six hundred and eighty-seven inches. These boxes can be set one upon another, and then have a space left between the potatoes, and are thus well adapted for use in storing a part of the crop, or a small quantity for domestic use. The barred sides and bottom secure abundant ventilation. The bins in the cellar should have a space of four inches between the end and the wall, and between the sides; this is easily made by placing a short rail between them, or a piece of four by four scantling, and this will relieve the sides from the bulging pressure of the potatoes. It is advisable to have a well-built root cellar, or a cellar under the barn, for storing potatoes; a house cellar should never be used for this purpose.

GRINDSTONES AND FRAMES.

A grindstone, to do good service, should be at least three feet in diameter and two and one-half to three inches in thickness, having a bevel on each side of the face for grinding on. It should be quite free from hard spots of iron pyrites, which are injurious to tools, although they may be .taken out with a sharp-pointed punch. If it is not centered truly it will work out of shape and soon require trueing up. It should run as fast

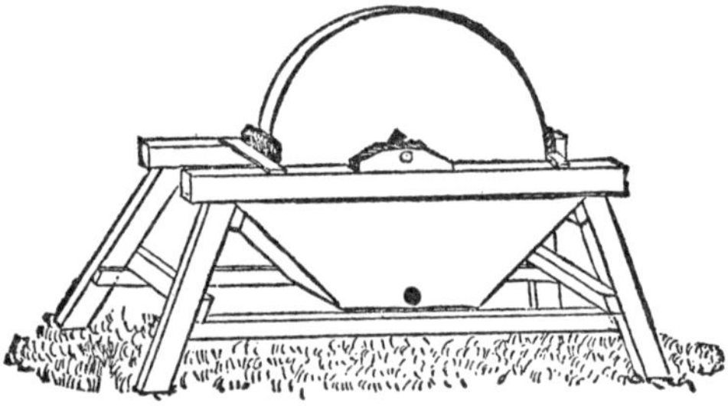

Fig. 70.—GRINDSTONE SET.

as possible, as it does its work better and more quickly. To prevent it from throwing water, a piece of bagging should be fastened to a staple fixed across the frame on each end (as shown in figure 70), but not so close as to grind it out; this will catch the excess of water and yet keep the stone wet enough and clean it. The stone should be kept in the shade and never in water, which softens it and makes one side wear faster than the other. The water box should have a hole in it to let out the water and keep the stone dry when not in use. In grinding, it should mostly turn from the tool, and if used otherwise, great care should be taken by the one who holds the tool, not to gouge the stone.

Figure 71 shows a novel style of frame for a grindstone. The frame proper consists of the iron part or bearing of a reaper reel. The arms to which the reel sticks were fastened, are all broken off but one. To this one the crank is bolted, as seen in figure 71. Four holes are drilled through the rim of the reel-wheel, to which

Fig. 71.—GRINDSTONE FRAME.

is bolted a hard wood board one inch thick, and having a square hole half way through, in which the center block fits. A bolt passes through a board block to a strip of iron, which may be bent to form a crank for

Fig. 72.—CROSS-SECTION. Fig. 73.—CENTER BLOCK.

foot-power. A cross-section of the stone as hung is shown in figure 72; the center block and board to which it is fastened are seen in figure 73. This frame should be bolted to a post or tree.

TOOL HOLDER.

Many a boy, and his father as well, who has toiled over the grindstone to sharpen tools, will be pleased with the device shown in figure 74, for giving a smooth, even edge to tools, which can be held by the hands, while the stone is turned by a treadle or a horse-power. It is a triangle of wooden bars, put together as shown, having a sharp pin at the point, a clamp for holding the tool at the center, and holes at the sides for tying an axle helve with cords, to keep it firm. The grindstone is near a wall or a post,

Fig. 74.—TOOL HOLDER.

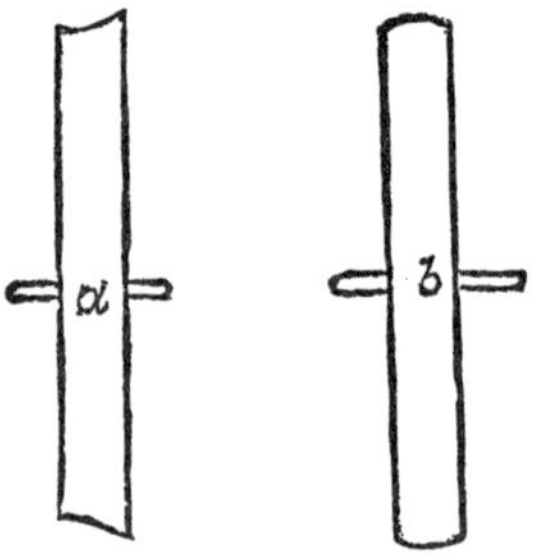

Fig. 75.—DAMAGED AND REPAIRED GRINDSTONE.

and the pin is pushed into this to hold the frame. The frame is then held in its proper position by the hands, and if held firmly, will grind an even bevel on any tool. A scythe, or a cutting-bar of a mower or reaper, or a chisel, can thus be ground perfectly and with little labor.

HOW TO REPAIR A GRINDSTONE.

Usually a grindstone is worn out of level, and very irregularly. This is scarcely to be avoided when such a large variety of tools, including scythes, mower sections, axes, hoes, and many other tools are ground. After ordinary use, those who are not careful to preserve the stone true, with smooth and slightly rounded face, the stone appears as at *a*, in figure 75. It is then beyond the power of the owner to repair the damage, unless he is an expert mechanic, when he takes a piece of old stove-plate and grinds the stone down to a slightly rounded or beveled face, like that shown at *b*. The best way to do this is to take a spade or a shovel, and turning it back upwards, to grind it sharp against the turning of the stone. This will bring the stone into the right shape, and in sharpening the spade, do a useful job at the same time.

A WOODEN MANGER FORK.

The common method of pitching fodder into mangers with a steel-tined fork, is often accompanied with harm

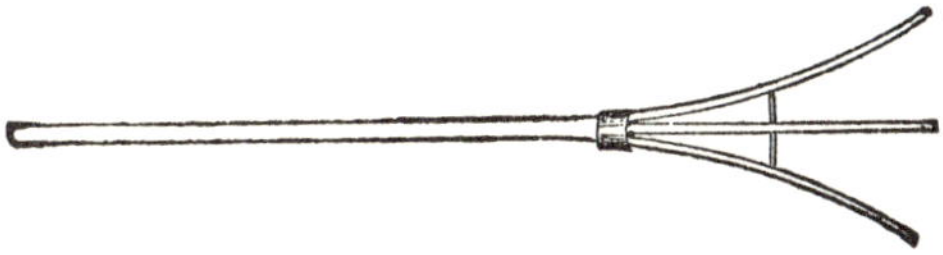

Fig. 76.—A MANGER FORK.

to animals. They will crowd around the rack or manger, and frequently receive an accidental thrust in the head or body with the sharp fork. Not infrequently an eye is lost, and with a horse this is a serious matter. The wooden manger fork shown in figure 76 avoids this danger. It is made of a piece of hickory or oak six feet long, an inch and a half wide, and an inch thick.

Four feet of its length is shaped round for a handle. The other end is sawed or split into three equal parts, to within a few inches of the rounded portion, where an iron band is placed. The "tines" are spread apart, and held in position by a wooden brace placed between them.

The tines are rounded, smoothed, and slightly sharpened at their points.

HOME-MADE AND USEFUL CHAFF FORKS.

Figure 77 represents a home-made fork with tines about two feet long, and having a spread of twenty inches. The teeth are straight above, and curved towards the point. They are fastened by screws to the three-inch

Fig. 77.—LARGE CHAFF FORK.

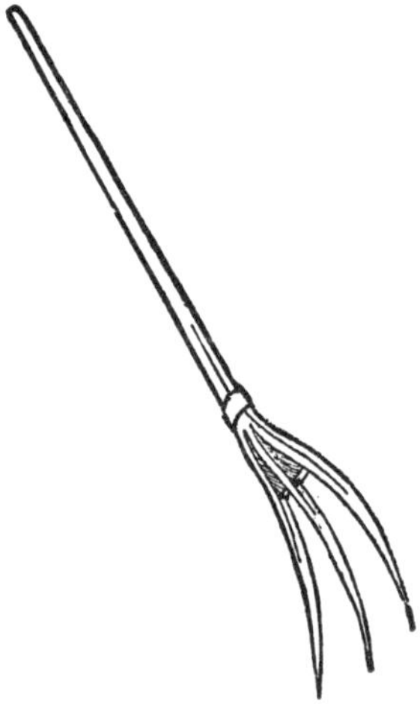

Fig. 78.—SIMPLE CHAFF FORK.

hard-wood head, and strengthened by an iron rod near the head, and by a round wooden rod, which passes through them and to which they are tacked fast. The head is strengthened by a similar piece of oak or ash, half an inch thick, screwed upon its edge, and through which the handle passes. This is of ash or hickory, large enough around to give the hand a good hold, and is fastened by wiring to the top side of the head. Such a fork may be made quite light, and the six tines being only four inches apart, will handle either chaff or light straw to good advantage. We give, by way of comparison, a simple chaff fork, figure 78, made by first binding and then carefully splitting a single piece of hickory or ash, handles and tines being formed of the same stick. A ring-ferule or band of wire is placed at the point beyond which the splits may not go, and after the splits are made, the tines are spread apart by wedge-shaped pieces of wood. These forks are easily made and are the very best stable forks that can be used. There is no danger of pricking horses or cattle with them, and if one be carelessly left in the stable, or falls down, neither man nor beast is likely to be hurt.

STABLE SCRAPER AND BROOM.

The manure gutter is easily cleaned out with the scraper and broom shown below. The scraper, figure 79, is made

Fig. 79.

Fig. 80.

to fit the width of the gutter, and brings the manure to the trap-door. The broom, figure 80, is then used to sweep the waste matter from the floors into the gutter, and from the gutters into the trap-doors, leaving the floor clean and clear for a new supply of litter.

A STRAW OR HAY HOOK.

A convenient hook for pulling straw or hay out of a stack for distribution among sheep or cattle, is shown in figure 81. It consists of a stout pole pointed at one end; a slit is cut through it and a hook is pivoted as shown in

Fig. 81.—STRAW OR HAY HOOK.

the engraving, so that it will be pushed back when it is thrust into the stack, and drawn forward, when it is pulled out. A strong cord helps to strengthen the hook, When the hook is pulled out of the stack, it brings a quantity of straw or hay with it.

FORK FOR HANDLING STONES.

The fork, figure 82, for lifting stones will prevent many a back-ache. It should have four prongs, which are

Fig. 82.

curved so as to hold the stones, and a strong handle. By a knack in giving a quick jerk, a heavy stone can be lifted and thrown into a wagon, and without stooping. Having used one of these contrivances to pick up stones, we can speak with knowledge of its usefulness. It is made of prongs of horse-shoe iron, welded to a heavier cross-bar, which has two strong straps to receive the handle.

SALT BOX FOR STOCK.

Salt should be given regularly to horses, cattle, and sheep, but it is rarely so given, because a supply is not kept handy for use. The box shown in figure 83 may

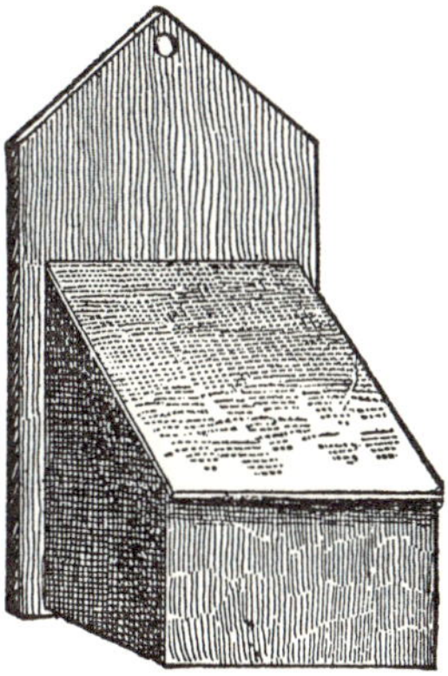

Fig. 83.

be hung in a stable or shed, or to a tree or post in the pasture. The salt is protected from the rain, and if replenished when necessary, the stock will be supplied with it regularly.

SAFETY SINGLE-TREE.

In plowing among fruit trees or in corn, single-trees having the traces arranged the usual way, will do much injury to the trees or corn. There is a method in arranging the traces which will avoid all this, as can be seen in the illustration, figure 84. A knot is made on the end of the trace rope, when the rope is passed through the hole made for the purpose, and brought around in the grooved end of the single-tree. To prevent the rope from getting out of place, it is wired or tied with strong cord. If the tree is struck by the end of the single-tree, thus guarded, it slides off without doing much injury. If the trees are young and small, with smooth and tender

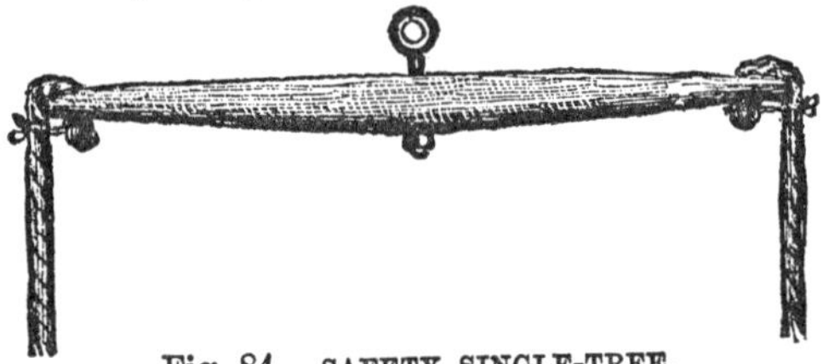

Fig. 84.—SAFETY SINGLE-TREE.

bark, it is well to wrap the end of the traces, for about eighteen inches from the single-tree, with old cloth, to prevent the rough, twisted rope from chafing the trees. Always use a shorter single-tree in plowing and cultivating an orchard than in ordinary plowing, and also use a small horse or mule to do the work, as this allows of more thorough work, and with less liability of injury to the low branches or the trunks of the trees.

ROOT PULPERS AND CUTTERS.

Those who feed beets, turnips, carrots and other roots, find it necessary to reduce them by some cheaper method than cutting by hand with a knife. An excellent machine for pulping roots is shown in figure 85. It may be made by any carpenter in two days, at a cost of about six dollars. The plan of the machine is given in the engraving. It is simply a square or oblong box, with a spiked cylinder

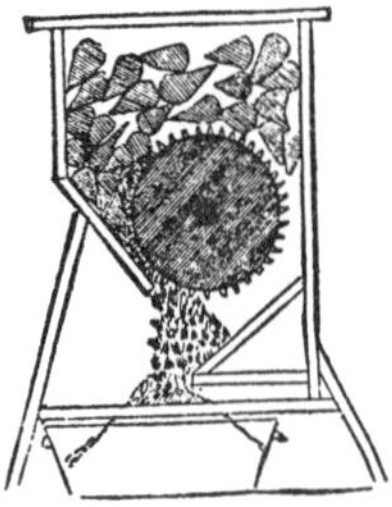

Fig. 85.—ROOT PULPER.

fitted in it, the cylinder having a square gudgeon at one end, to which a handle is fitted. To save expense the heavy wheel and handle attached, of a fodder cutter, may be taken off and used on the root pulper, as the two will rarely be used at the same time. The cylinder is closely studded with sharp, chisel-pointed spikes. These teeth are made of one-quarter inch square bar iron, and are three inches long; the sharp edges are worked out on an anvil, and are chilled by immersion in

Fig. 86.—HOME-MADE ROOT-CUTTER.

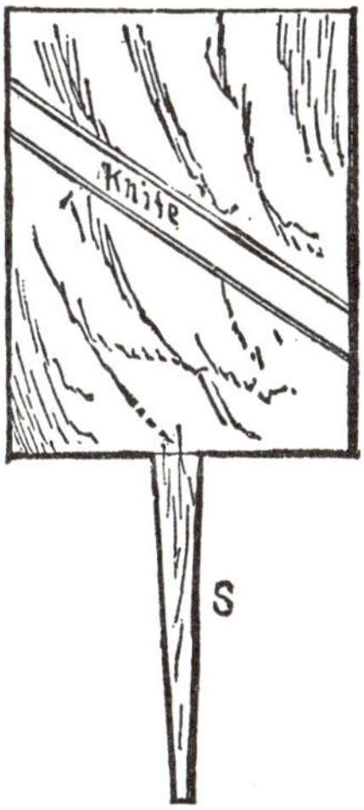

Fig. 87.—ROOT-CUTTER SLIDE.

cold salt water when red hot, the other end being cut with a screw thread. To secure strength, and to make the machine work with more ease, the cutters are screwed in so far as to leave only half an inch or a little more projecting. A still cheaper form is illustrated in figure 86. At *A* is seen the hopper which is without a bottom. The slide, figure 87, contains a two-edged knife, and runs in the grooves, *G G*, in the top of the frame, close to the bottom of the hopper. Near the bottom of the frame is a roller, *R*, into which is fitted the handle, *H*. This is connected with the slide by the rod, *R*. The knife shoùld be about four inches wide and one-quarter inch thick, be placed diagonally in the slide, leaving half an inch space between it and the bottom of the slide. When using the apparatus all that is necessary is to move the handle to and from the hopper. It works easily and quickly, is durable, and with fair usuage is not likely to get out of order. A ready way of chopping a few roots, is to use a spade ground to a sharp edge, and a box in which the roots are quickly reduced to slices. A basket of turnips or apples, can be sliced in this way in one minute. For a larger quantity, a chopper may be made as in the engraving, figure 88. It has two long blades. and the roots are hashed up rapidly, and all danger of choking is avoided. A common cast-iron winged gudgeon, having steel strips riveted on the edges, answers as well as one forged out by a blacksmith, at several times its cost. The roots, so cut, may be mixed with meal, and fed to the cows. Apples are excellent for dairy cows when fed in this way, and largely increase the flow of milk, besides being healthy for them.

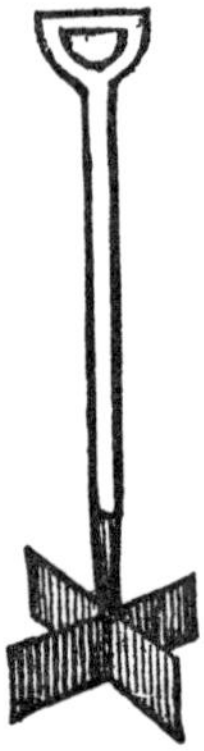

Fig. 88. ROOT CUTTER.

ROOT WASHERS.

A convenient wäsher for potatoes and roots, consists of a kerosene barrel hung in a frame, as shown in figure 89, on next page. Two openings are made in one side of the barrel—a large one, two staves wide, and a small one only one inch wide. The pieces cut out are used for lids, both of which are fastened with hinges and buttons, and are made to fit tight by having thick cloth tacked around their edges. A bushel of potatoes or roots are placed in the barrel, with two or three buckets of water, the lids are closed and buttoned, and the barrel is slowly turned. If they are very dirty, open the small aperture, and by turning the barrel back and forth allow the water and mud to run out. Add clean water and

Fig. 89.—ROOT WASHER.

turn again. They will soon be cleansed, when the large aperture may be opened, and the roots or tubers emptied into a basket. The fastenings at each end of the barrel can be made by any blacksmith, and they should be bolted on with one-quarter or three-eighths inch bolts. With this simple contrivance a man can wash a large

Fig. 90.—VEGETABLE WASHER.

quantity of roots in a day without catching cold or a chill. If kept out of the sun, such a contrivance will last a lifetime. In figure 90 is shown a potato and vegetable washer for household use. The ends of the cylinders are cut out of inch board and are twelve inches in diameter. The shaft runs through and has collars, to which the ends of the cylinders are fastened to hold them firm. Strong, tinned wires are fastened from end to end, as seen in the engraving. Five of these are fastened together, and form the lid to the aperture through which articles are admitted. The end of the lid is fastened by means of a loop, which springs over a button. The vegetables to be washed are placed in the cylinder, the box is half filled with water, and by turning

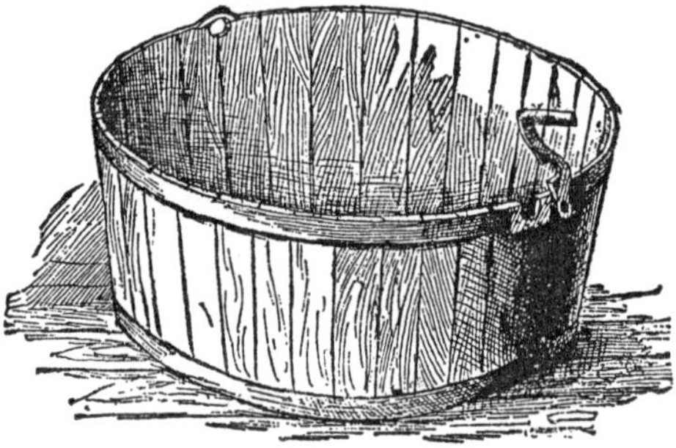

Fig. 91.—TUB FOR WASHER.

the crank, or by moving it back and forth, they are quickly cleansed. Narrow wooden slats may be used instead of wire, if desired. An ordinary tub, or a half barrel, arranged as seen in figure 91, may be used instead of the box.

CLAMPS AND STOOL FOR REPAIRING HARNESS.

The device shown in figure 92 combines a stool and a clamp for holding harness work. The bench or stool, *b*, of any desired size, is supported by two legs near one end. The other end is held up by the foot of the long claw, extending to a convenient height for the operator. A shorter claw, *c*, is fastened to it by a cross-piece, *p*, about an inch thick and three inches wide, passing through a slot in the jaws, in which it works easily but firmly on two iron pins, a little more than half-way up from the bench. In the lower end of the short jaw an eccentric works on a pivot and against a projection on the larger

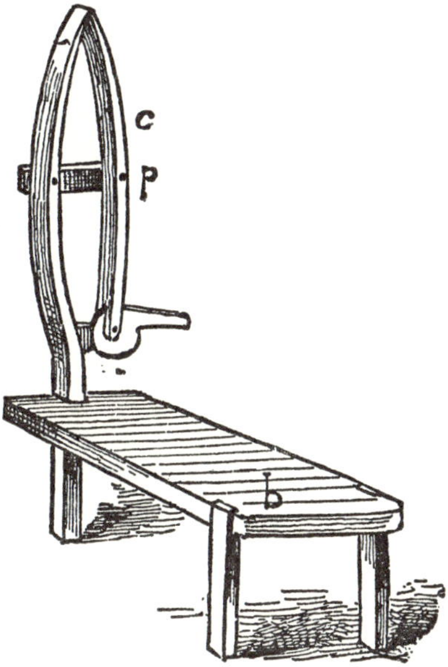

Fig. 92.—A HARNESS STOOL AND CLAMP.

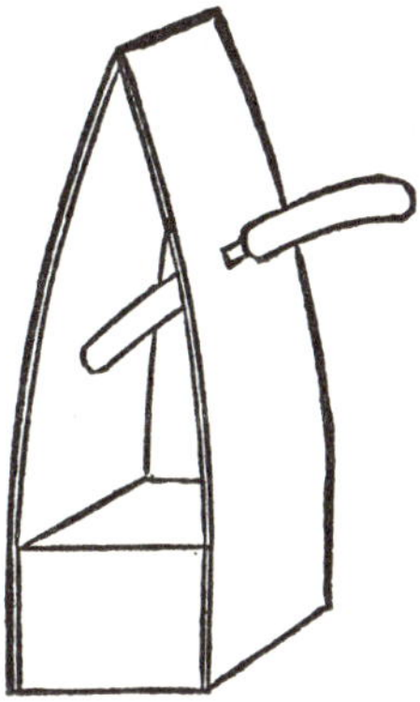

Fig. 93.
HARNESS HOLDER.

jaw. Depressing the handle to this eccentric or cam, closes the jaws at the top with all the force desired.

A simple holder without the stool is shown in figure 93. Two staves of a flour barrel are sawed off at a convenient length for holding between the knees, while sitting on a chair. The sawed ends of the pieces are securely nailed to the opposite sides of a block of wood. A hole is cut through the middle of one side piece, in which a lever is placed for opening and closing the holder. The lever may be readily made of such shape that it will always remain in the hole, ready for use.

The curves of the staves will furnish sufficient spring to hold the harness.

A BOX SAW-HORSE.

The novel saw-horse shown in figure 94 is made of a dry-goods box, of inch pine boards, thirteen inches long, eighteen inches wide, and twenty-four inches in height. Upon the outside of one end are nailed two cleats, and on the inner side three cleats, the position of which is

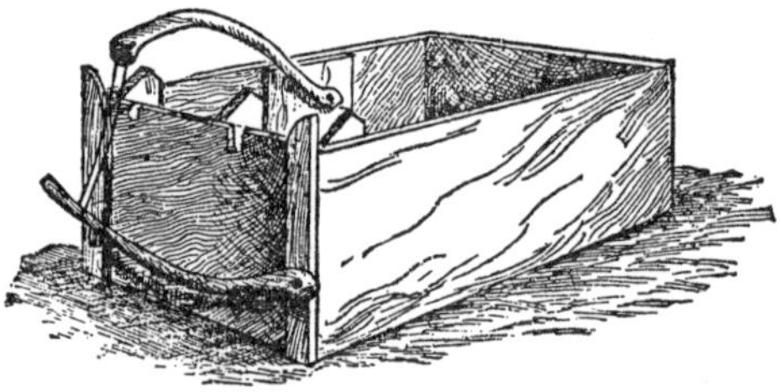

Fig. 94.—NOVEL SAW-HORSE.

shown in figure 94. The curved lever above the box is intended to do the hard work usually imposed upon the sawyer's left knee, viz., holding the stick sawed in place. The necessary pressure of the lever is effected by means of the treadle and the small rope or sash cord connecting the two. The lever should be so attached to the side of the box that the loose or curved end rests upon the stick, held in place by it, about midway between the center and left diagonal cleats. The treadle should extend, when horizontal, eight inches beyond the left side of the box. In using the horse, raise the lever with the left hand, with the right place the stick to be sawed so that the point where it is to be cut is over the U; the lever is dropped or pulled down upon the stick; the left foot is placed upon the treadle; a slight pressure will hold the stick securely. The sawyer, thus using both limbs for support, and standing nearly erect, will find wood sawing an easy though vigorous exercise, quite exempt from many of the old-time aches and pains. If the horse is to be used in a wood-house—a room having a floor—it is well to secure it by screws to the floor; if out of doors, it may be ballasted with a few bricks or stones, or be fastened to a frame.

LONG SAW-BUCKS.

In cutting fire-wood from long timber or sawing lumber, it is convenient to have a long saw-horse. Two patterns are illustrated herewith. To make the one shown in figure 95,an oak stick averaging half a foot in

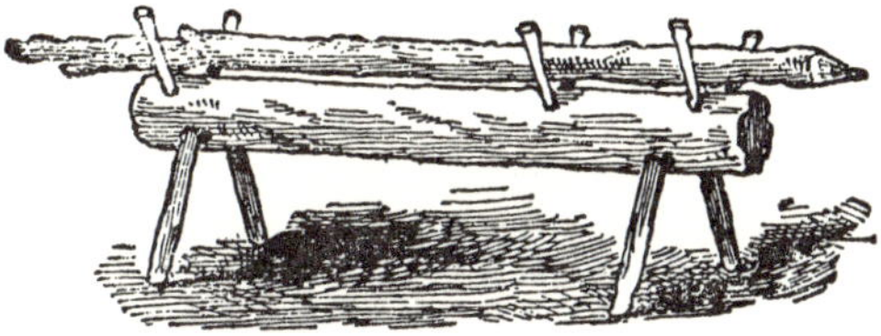

Fig. 95.—A LONG SAW-BUCK.

diameter, was selected from the wood pile, and a piece five foot long cut off. Two one-and-a-half-inch auger holes were bored near each end, not quite opposite each other, to avoid weakening the timber at one point, and

Fig. 96.—A LIGHTER HORSE.

four strong sticks from the same wood pile were driven in for legs—a little under two feet long, and standing well slanting outward. Six one-inch auger holes were bored in the top, and split-out pegs, eight or ten inches long, were driven in, in a position to firmly hold the wood to be sawed. The two pegs of each pair are not directly opposite, but separated far enough for the saw-cut to run down between them. Of the first pair one is four inches from the end, and the other seven inches back. The second pair is fifteen inches back of these, and the other in the farther end of the horse, these last answering as a support to the long end of the wood to be cut, the other two pairs being used as the saw-horse. When a stick is reduced to five feet or so in length, it is drawn forward and wholly supported on the two pairs of pins nearest together.

The other horse, shown in figure 96, consists of an ordinary saw-horse having a block nailed across its legs on one side, forming a rest for the end of a long stick, which at the other end is fastened into half a saw-horse, *a*. The piece to be sawed is laid on the three rests thus formed, the end to be sawed being placed at *a*. As each length is sawed off, *a* is shoved toward *b*, the proper distance. It will be seen that this saw-horse can be lengthened out or shortened up, to suit the length of the stick.

HOW TO TIE A BAG.

Figure 97 shows a simple and easily made bag-tie which effectually prevents any slipping, if properly ad-

Fig. 97.—BAG TIE.

justed. Take any strong cord about eighteen inches long and double it as herewith seen, passing the ends through, making a loop around the mouth of the bag. Now pull as tightly as possible ; then take an end of the string in each hand and pull again in opposite directions; pass the string completely around, make a knot, and double or single bow-knot, and the work is done. A very little experience will make one expert, and he can then make sure the bag will not come untied.

A HOME-MADE RAKE HEAD.

Figure 98 shows the end view of a hand-rake with the tooth inserted. The head-block should be made of green hickory, free from knots and curls, while the rake-teeth must be of dry, well-seasoned oak or hickory, and having grooved places in them, as is seen in the illustration. The teeth are rounded and are driven snugly into the green head-piece, which latter, in drying, will hold the seasoned teeth so firmly as to effectually prevent them from ever coming out. In fact, so tightly will they be held, that they can scarcely be driven out when the head-block has become thoroughly seasoned, the shrinkage of the green wood acting as a permanent vise. The same principle might be utilized in other small implements.

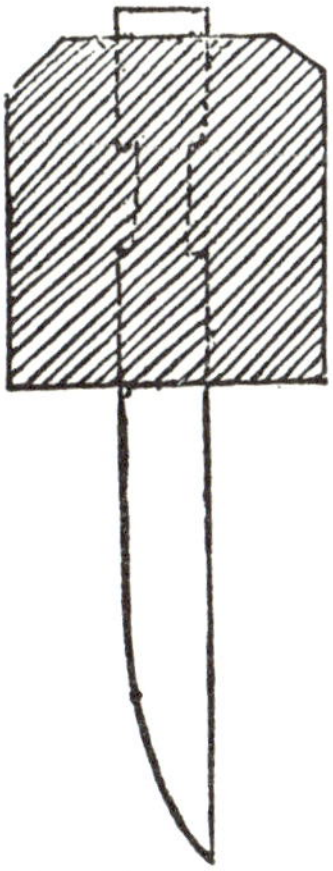

Fig. 98.—A DURABLE RAKE.

WORKING BUILDING STONE.

Stone is the most durable and the cheapest building material where it is plentiful on the farm. By a little management the stone can be brought to a convenient shape for use. The tools required, shown in figure 99, are: a chipping hammer, a wedge and steel feathers, a striking hammer, drill and a bar for opening cracks in the stone. The clipping hammer has a broad, sharp edge, and acts as a chisel for dressing the faces ; and the sharp edges of the rectangular head, two by four inches, serve

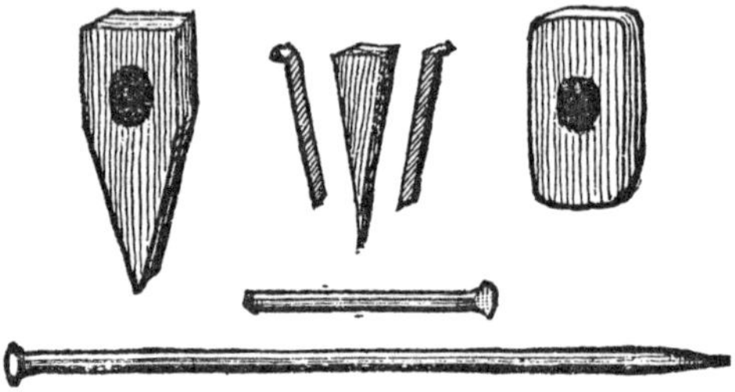

Fig. 99.—TOOLS FOR STONE WORK.

to dress down the edges and corners of the stones. The wedge is three by one and a half inches, and the feathers are plates of steel as wide as the wedge, which they serve to protect. The striking hammer is three inches square, and six inches long, with a beveled edge around the face. The drill is of one and a quarter inch octagonal steel, and is eighteen inches long, or if there be two, one is twelve inches long. The bar is four and a half feet long, and has a sharp-edged steel point for striking into cracks and splitting the stone, which it is usually easy to

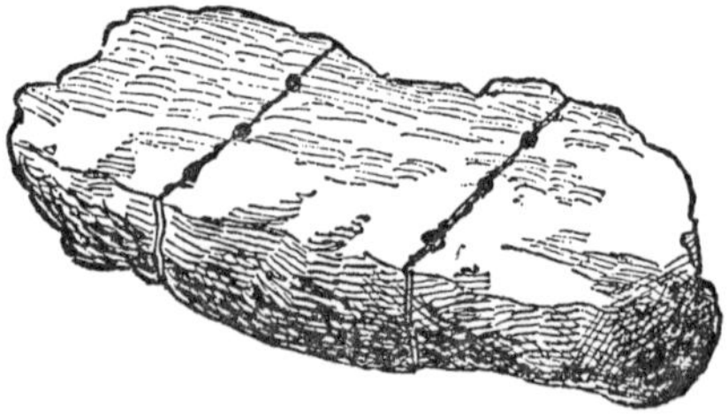

Fig. 100.—BREAKING A LARGE STONE.

do. A large stone is broken by drilling a few holes in it with a one-inch drill, and chipping a groove across the face along the line of holes, as shown in figure 100. Small round wedges, with small feathers, are placed in each hole, and they are struck one after the other, in rotation.

By this method very large blocks are split with an even face. A small stone is easily split by chipping grooves across it, and then repeatedly striking along upon the groove with the face of the hammer.

BLOCK FOR SAND-PAPER.

Sand-paper is put up by the manufacturers in quires of sheets nine by eleven inches in size. As used by many workmen, nearly a fourth of each sheet is wasted by folding and crumpling over improperly shaped blocks. A convenient block, figure 101, which permits the use

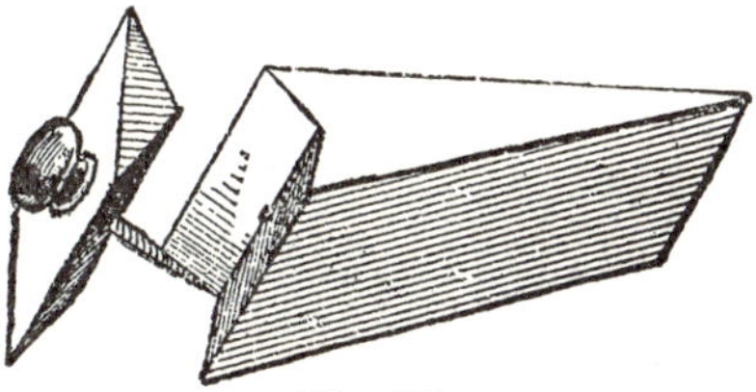

Fig. 101.

of all the sand-paper, is here described. Make a wedge-shaped piece of hard wood, one and a half inch thick, three inches wide, and five and one-quarter inches long, tapering from the head to a sharp edge. Cut a V shaped hollow across the head. Fit a piece three inches

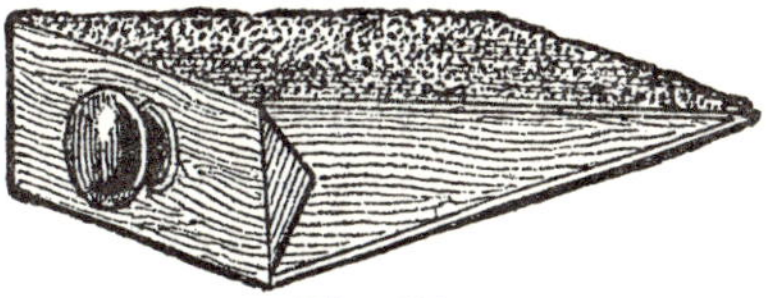

Fig. 102.

long, of hard-wood, exactly to this hollow. Insert in the head a wood or porcelain drawer knob seven-eighths of an inch in diameter, fastening it securely by a long screw. Cut a sheet of sand-paper into three equal parts, three by eleven inches. Fold one-fourth of an inch at each end of a strip of sand-paper, and slip under the head-piece by loosening the screw. Tightening it will hold the paper fast and smooth for work. A common wood screw may be used in place of the knob, but is not as convenient, as it must be turned by a screw-driver.

CHAPTER IV.

APPLIANCES FOR THE BARN, PASTURE AND DAIRY.

CONVENIENT STABLE VENTILATOR.

It must not be supposed that fresh air in the winter is to be excluded from stables for the purpose of keeping the animals warm. Warmth alone is not comfort. An animal may suffer from cold in a close, damp, impure air, which is really warm, while it will be quite comfort-

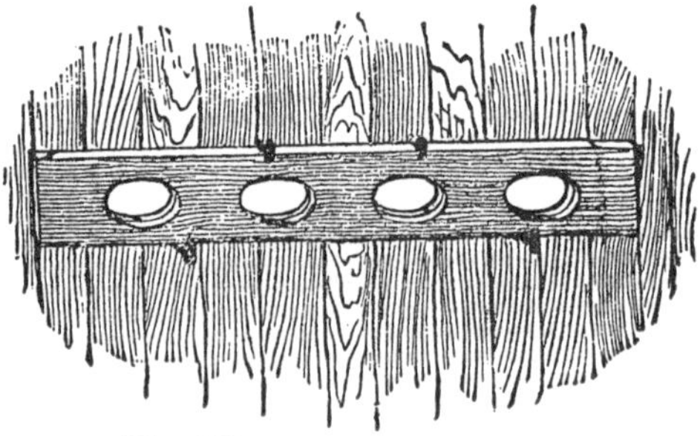

Fig. 103.—STABLE VENTILATOR.

able in fresh, pure air, which is much below freezing temperature. The absence of oxygen in the one case reduces the vital warmth, while its abundance in the other case maintains an agreeable and comfortable feeling. Consequently, ventilation of stables is necessary, even in the coldest weather, to keep the animals in good health and in comfort. But it should be regulated judiciously by a suitable provision of openings at the upper part of the stable, and these should be made so as to be readily opened and closed. A row of holes cut in the wall near the upper floor, figure 103, and covered with a sliding-board, having precisely the same kind and number of holes to match those in the wall, will afford suitable ventilation for the stable at all seasons. The covering board slides back and forth on the pins shown, and covers or opens the holes as may be desirable, according to the condition of the weather.

LIGHT NEEDED IN BARNS.

Here and there we see an old-style barn, built by our grandfathers, the only window being a single row of panes over the large door. Through this comes all of the light admitted to the barn, except what comes through the open cracks between the boards. When a barn of this kind is filled with hay it is comfortably warm, but very dark; by midwinter the hay, being half consumed, leaves the walls unprotected. With the light come in also the cold wintry winds to chill the cattle. Our fathers built some barns warmer, covering the walls with shingles or the cracks with narrow battens. The light being thus shut out, it was necessary to have windows; so they put in just enough to enable them to see to feed their cattle. It was left for our generation to build barns that are tight, comfortable, and well-lighted. But even at present many farmers do not realize the importance of light in a cattle barn. Experiments show that a herd of milch cows not only keep in better health and condition by having plenty of light, but they give more milk. Every barn should be provided with abundant light and sunshine on the side where the cattle stand. The practice, which is far too prevalent, of keeping cows in a dark and damp basement is not a good one. They can not have the sunshine and pure air so necessary for good health. Windows that are exposed may be protected for a trifling sum by covering them with wire netting. The day of windowless barns has passed; but some of our new barns would be improved by a few more windows.

LANTERNS IN THE BARN.

It is estimated that nine-tenths of all fires are caused by carelessness. Never light a lamp or lantern of any kind in a barn. Smokers may include their pipes and cigars in the above. The lantern should be lighted in the house or some out-building, where no combustibles are stored. A lantern which does not burn well, should never be put in order in the hay mow. There is a great temptation to strike a match and re-light an extinguished lantern, wherever it may be. It is best to even feel one's way out to a safe place, than to run any risks. If the light is not kept in the hand, it should be hung up. Provide hooks in the various rooms where the lights are used. A wire running the whole length of the horse stable, at the rear of the stalls, and furnished with a sliding hook, is very convenient for night work with the horses. Some farmers are so careless, as to keep the lamp oil in the barn, and fill the lantern there, while the wick is burning. Such risks are too great, even if the buildings are insured.

SAFETY STICK FOR MARE'S HALTER.

Figure 104 shows a simple method of preventing colts from getting tangled in the mare's halter, and so becoming strangled. A piece of wood, eighteen inches long and two inches wide, is bored with a half-inch hole at each end, as shown in the engraving. The halter is passed through these holes and fastened in the usual

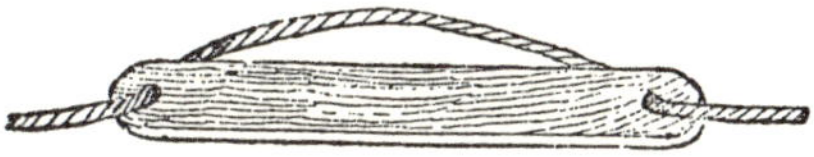

Fig. 104.—HALTER STICK FOR MARE WITH FOAL AT FOOT.

manner. The wood should be of tough oak or hickory, so that it will not break. It entirely prevents the formation of loops in the halter.

TO KEEP A HORSE FROM JUMPING.

Figure 105 shows a hopple to restrain a horse from jumping. It consists of a surcingle about the body of the horse, together with two short straps that pass through

Fig. 105.—HOPPLE FOR A HORSE.

the surcingle and around each foreleg, being buckled so that when the horse stands upright, the strap will fall about half-way to the knees. This arrangement, which allows the horse to walk quite freely, prevents its running as well as jumping. A similar plan is to connect the forelegs of a horse by straps secured just above the knee, but those who have tried both plans prefer the one herewith illustrated. Some horses are difficult to catch when at pasture, and this device will prove valuable in such cases.

COUPLING HORSES IN THE PASTURE.

Cut a piece of tough wood two feet six inches long, two inches in diameter; shave off the bark and bore a three-quarter or a one-inch hole near each end, tie a piece of half-inch rope around each animal's neck,

Fig. 106.—COUPLING FOR HORSES.

making a loose collar that will not slip over his head; take a loop of the rope and pass it through a hole in the bar, and into the loop insert the key, made of a piece of a half-inch oak board, two by three inches, shaped as in figure 106. The board being rounded at the top, will allow the rope to turn easily in the yoke and prevent choking. Always couple the animal that is likely to stray with the one that is not. This contrivance, used frequently on Southern and Western ranges, is approved by some as safe and convenient, and condemned by others as dangerous and uncomfortable for the animals. We give the illustration and description for what they are worth.

A SIMPLE TETHER.

Figure 107 shows a tether for a horse or cow which obviates the danger of an animal becoming entangled as when staked out in the usual way. It is made as follows: Take a stout piece of timber, *a*, *b*, three and one-half feet

Fig. 107.—TETHER FOR HORSE.

long; fasten a ring at *a*, and one at *c*, six inches from the lower end. Take a pole, *c*, *d*, making it long enough to extend back of the animal's heels three or four feet, and fasten a ring to each end. An iron spike, *f*, with a ring, *e*, in the end, is driven in the ground. The irregular line represents a cord of wire of any desired length.

Fasten the rod, *a*, *b*, to the halter at *a*, with a leather strap, also *c*, *d* to *a*, *b*, in the same way at *c*; tie one end of the cord in the ring at *d*, and the other in the ring in the end of the iron spike. The ring at *c*, six inches from the lower end, prevents taking up the cord, and thus entangling the animal. The end, *b*, will slide over it as the animal grazes.

CHAIN CATTLE TIE.

Various methods have been devised for coupling cattle in their stalls in a more humane manner than by stanchions. The common chain tie passes about the animal's neck, and slides up or down upon a post or iron rod, attached to the stall or manger. The tie, figure 108, is similar, except that the neck-chain is connected with two posts or rods, upon which it slides. The improvement consists in using rings upon the posts, and con-

Fig. 108.—AN IMPROVED TIE FOR CATTLE.

necting the side-chain with the neck chain by means of snap-hooks, attached to the central ring as shown in the engraving. This enables one to adjust the tie to any width of stall, say from three to four feet, and have it reasonably taut. The advantage of this method of fastening cattle over any other is, that while great freedom is given the head, so that a cow can lick both sides and lie down with her head upon either side, she has no more backward and forward motion than if she stood in stanchions, hence must leave her droppings in the gutter—if the stall is of the proper length. There is a constant tendency to give cow stalls too long a floor. Every cow should lie with her rump four to eight inches beyond the floor. The only objection to this is that the cows' tail will sometimes become wet from lying in the gutter. If, however, this is given a pretty sharp fall and considerable breadth, water will not accumulate, and there will be no inconvenience experienced on this score.

AN UNPATENTED CALF FEEDER.

Undoubtedly calves which take nourishment directly from the cow, do better than those which take it from the pail, unless care is taken to feed them slowly. An

Fig. 109.—CALF FEEDER.

artificial udder is shown in figure 109, made of strong water-proof duck in the shape of a cow's udder, and furnished with teats, each filled with a piece of sponge. The mouth of the bag may be closed by means of clamps, figure 110, and the bag hung up in the calf pen. The calf will get its milk slowly and along with plenty of

Fig. 110.—CLAMPS FOR CALF FEEDER.

saliva, which is an indispensable aid to digestion. It is the want of an adequate quantity of saliva with the milk, which causes so much indigestion in calves that are allowed to drink milk from a pail.

TWO KINDS OF MILKING STOOLS.

The construction of a very good milking stool is readily seen in figure 111. Upon a hard-wood board,

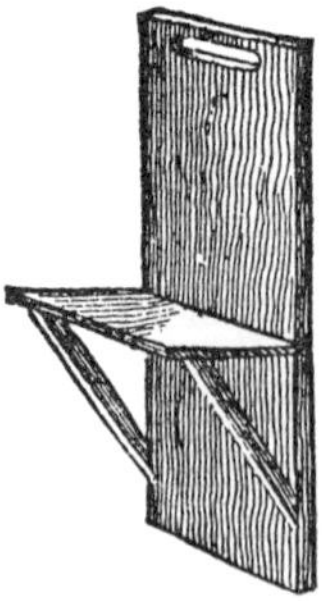

Fig. 111.—MILKING STOOL.

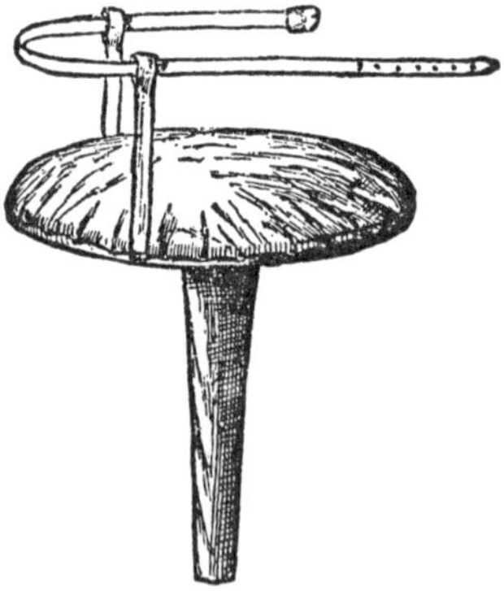

Fig. 112.—MILKING STOOL.

twelve inches wide, one inch thick, and thirty inches long, fasten at right angles a board to serve as a rest. This should be eight inches wide, and as long as the width of the back-board. Strengthen the seat with stout braces. Cut a narrow opening in the long board, to admit the fingers, by which to carry the stool, or hang it up when not in use.

The other stool, figure 112, is designed for a man who has a good many cows to milk, and desires to carry his stool around with him, while his hands are left free. The seat consists of the bottom of a peach basket ; the single leg is made of a round piece of wood securely fastened to the center of the seat. The latter may be padded and covered as one chooses. Leather straps to reach up and around the waist of the milker, as shown in the illustration, should be firmly attached to the seat.

VAT FOR DEEP SETTING MILK.

The advantages of the deep setting of milk at a low temperature can be enjoyed by means of the simple cooler, figure 113. To make the cooler take six pine planks, two inches thick, twelve inches wide, and six

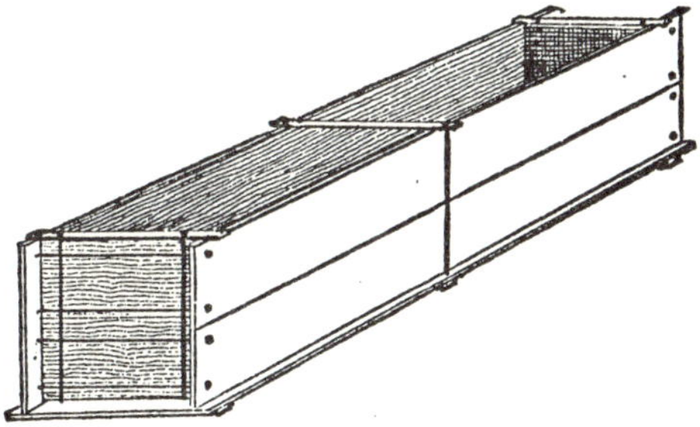

Fig. 113.—A COOLER FOR SETTING MILK.

feet long, four boards sixteen and a half inches long and twelve inches wide, and construct a box with the ends gained in with a groove a quarter of an inch deep. Place a rubber strip between the boards, and clamp with rods and bolts, to make it as tight as possible. Provide a lid to keep out dust and to shade from the sun. Place a faucet at the bottom, by which to run off the water when it has become warm. Set the cooler near the well whence cold water can easily be drawn, and keep the cans of milk submerged in the water. If there is a supply of ice, the temperature of the water may be still further reduced, and the cooler rendered more efficient. A box of the size given above will have room for twelve three gallon cans.

HOME-MADE BUTTER-WORKER.

The butter worker, figure 114, is made to stand upon a table or low bench, or when of large size, upon the floor. The lever works upon a rod and can be moved sidewise, an arrangement which we have seen in no other butter-

Fig. 114.—A SERVICEABLE BUTTER-WORKER.

worker, but which is a very desirable one. The table slopes forward, and has several grooves to carry the liquid down to a pail or a dish placed to receive it. The lever at the under side is leveled to a round or sharp edge, as may be wished. The cost of the worker is a mere trifle; it should be made of maple, ash or chestnut.

A CONVENIENCE FOR FLY TIME.

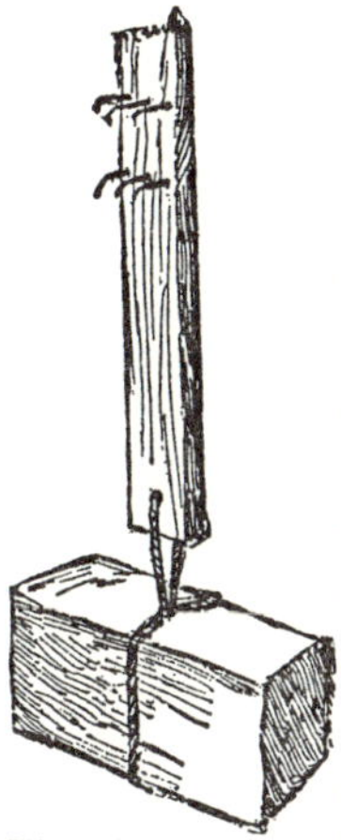

Fig. 115. — TAIL-HOLDER.

The comfort which a cow seems to derive from a free use of her tail during fly time, is not shared in any degree by the milker, and various means have been devised to hold the troublesome appendage in place. One of the latest is illustrated in figure 115. Half a dozen six-penny wire nails are driven through a piece of lath, and each point bent to a hook. A brick is suspended by a string from the lower end of the stick. As the milker sits down beside the cow, the hooks are thrust into the brush of her tail, leaving the brick resting in part on the ground or barn floor. After the first futile efforts to swing the brick by tail power, the cows learn to give it up, and the milker is free from a very great annoyance.

REINS FOR DRIVING OXEN.

Figure 116 shows a method of arranging the reins for a yoke of oxen. Each ox has a spring bull-ring placed in

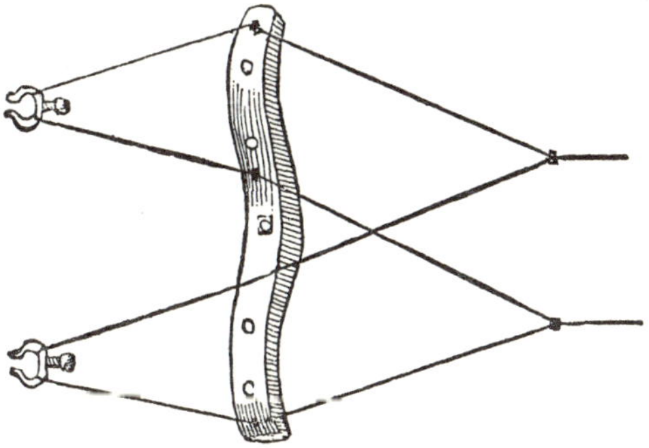

Fig. 116.—DRIVING OXEN WITH REINS.

his nose, and from these rings small ropes run back as seen in the illustration. Staples are driven into the top of the yoke through which the cords pass. It is claimed that with these reins a yoke of oxen can be guided and controlled with ease. The rings are quickly removed from the noses when work hours are over.

VAT FOR DIPPING SHEEP.

Sheep should be dipped twice a year. They suffer a great deal from vermin, which are destroyed by the dipping. After shearing, the ticks greatly annoy the lambs, upon which they gather from the shorn sheep and prevent

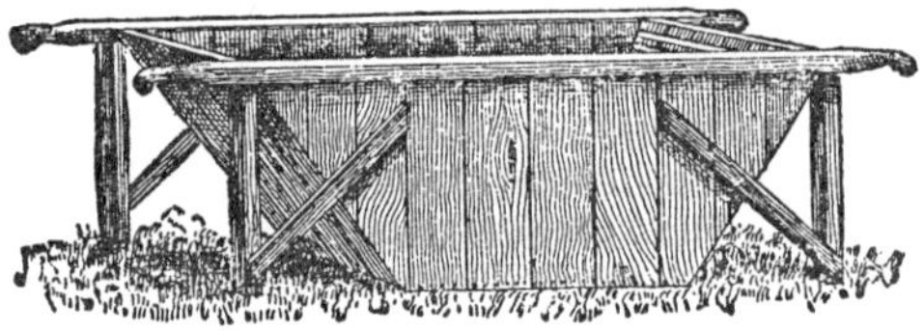

Fig. 117.—PORTABLE VAT.

their growth. The lambs, at least, should be dipped, to free them from these pests, but it is well to dip the whole flock, as a safeguard against the prevalent scab, and other skin diseases. A very good dipping vat is shown in figure 117. It is made of one and a quarter inch tongue and grooved boards, put together at the joints with pitch, and is furnished with handles, with which it can be moved from place to place. It may be six feet long, three feet wide, and three feet deep. The sloping ends have cleats nailed across them on the inside, by which the sheep are assisted to get out of the vat, upon a draining floor placed to receive them. A good dip is made of one pound of coarse tobacco, and one pound of sulphur, steeped in five gallons of boiling water. It is most effective when used at a temperature of one hundred and twenty degrees, and the sheep should be left in the dip long enough to have the wool saturated, and the skin well soaked by the fluid. A quantity of fresh dip should be kept in a boiler, to renew the old dip as it is diminished by use.

SHEEP-SHEARING BENCH.

Shearing benches will be found desirable, as they save the wearisome stooping over the sheep. A bench of this kind is shown in figure 118. It is made of stout strips nailed to curved cross-pieces. These are best bent by

Fig. 118.—SHEARING BENCH.

steaming them, or soaking them in hot water for some hours, or sponging them frequently beside a hot fire, by which the fiber is much softened and the wood is warped permanently. The legs are about twenty inches long. Any dust on the wool falls through the bars.

EAR TAG PUNCH FOR MARKING ANIMALS.

A punch, which is struck with a hammer, and even the new belt-punch pattern, now so generally used, inflict considerable pain; the blow in one case, and the very considerable pressure needful in the other, are both productive of suffering which can just as well be avoided. Some breeders have used with entire satisfaction a very simple contrivance, figure 119, which any machine shop can furnish from the engraving and description herewith. Take a piece of steel rod, say five inches long and about five-sixteenths of an inch in diameter. Fix this in a chuck and drill a hole endwise from *a* to *b* in the engraving, which leaves that end a hollow tube, the walls of which are about one-thirty-second of an inch thick, supposing a one-quarter inch drill has been used. Possibly a little smaller hole would be better. Then file a notch in one side at *b*, so that it will clear readily. Drill a small hole, *c*, near the one end, in which to fit a short piece of smaller wire, *d*, which forms a convenient gimlet-like handle. When finished, have it nicely filed to a taper at the hollow end, so as to form a thin cutting edge, which must be kept quite sharp. After being tempered it forms the best tool for its work ever invented. To use the punch, hold in the left hand a large cork, or a small block of wood, and carefully selecting the proper place between the ribs or ridges of the ear, press the punch snugly down, give it a quick, sharp twist, just as one would a gimlet, and the animal scarcely flinches at all, so slight is the pain.

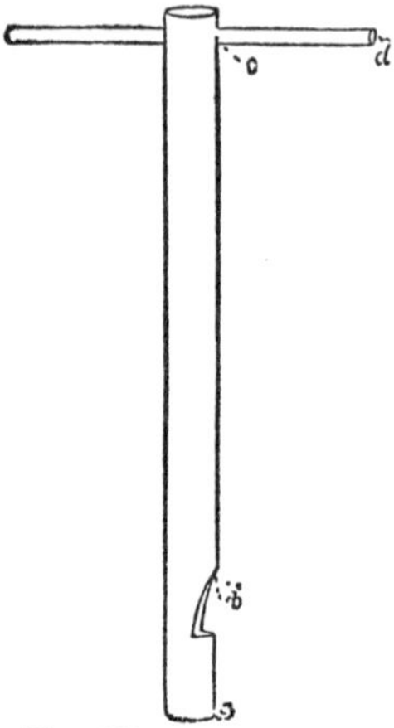

Fig. 119.—EAR PUNCH.

SEWING UP WOUNDS IN ANIMALS.

The winter season is always prolific of accidents, chiefly among horses, which are often badly blemished by cuts which are left to heal imperfectly, without any assistance.

When a horse with sharp calks kicks another, or when an animal falls upon ice, the skin is usually cut in an angu-

Fig. 120.—NEEDLE FOR SEWING UP WOUNDS.

lar shape and the flap of skin hangs over in an unsightly manner, or in a torn cut the skin gapes open and makes a wound difficult to heal. As a rule, a horse's wound

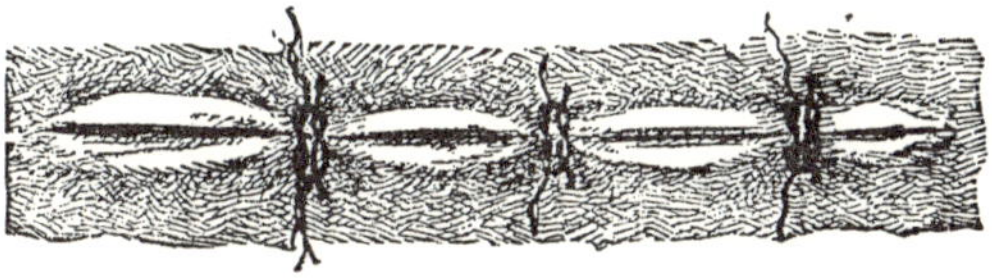

Fig. 121.—WOUND SEWED TOGETHER.

heals very rapidly under the simplest treatment. A curved needle, figure 120, is used to sew up severe wounds as shown in figure 121.

CHAPTER V.

WELLS, PUMPS, CISTERNS AND FILTERS.

WINDLASS AND TILTING BUCKET.

As ordinary pumps draw water only thirty-three feet perpendicularly, and practically only about thirty feet from the water surface, force-pumps or windlasses are required, for wells thirty or more feet deep. The common windlass with stop ratchet serves a fair purpose, but requires the bucket to be let all the way down by turning the crank backward. Various forms of brakes have been devised. Figure 122 shows the construction and operation of one. Two opposite corner pieces, *p*, extend six feet high above the platform, and a diagonal piece connect-

Fig. 122.—IMPROVED WINDLASS.

ing their upper ends supports a grooved pulley carrying the lifting rope. A hook, *h*, turning on a pivot, is thrown over the lever *b*, and slid along it far enough to hold the brake against the windlass firmly, when the hand is removed. A swinging iron rod catches in the small pin on the top of the bucket as it rises, and tips the water into the spout. With these little additions, a windlass and bucket are better than a pump, as the water is drawn fresh, with no tainting from the pump log. The actual force required to raise the same water is less with the windlass than with the pump, as less power is required than is wasted in the friction of the close fitting valves of the pump, and the friction of the water against the side of the tube.

WELL-CURB OF STAVES.

Figure 123 is a very strong and durable curb made of staves. A cooper can make it, setting up the staves, which are one and one-quarter inch thick, as for a barrel, using three iron hoops. The shaft of the windlass is also of iron, to which a wooden cylinder is fastened by a

Fig. 123.—A "BARREL" WELL-CURB.

couple of bolts driven through the wood and iron. In making the windlass, fashion the wood to the right size, and then split open the cylinder, cut a place for the shaft, fit it in, and then drive bands over the ends.

HEMLOCK FOR WELL-CURBS.

In many sections of country stone is scarce, and plank is used for curbing wells. Pine lumber gives a disagreeable taste to water. Hemlock lumber is usually cheaper than pine, and can be obtained at most lumber yards. Five hundred feet of lumber are sufficient for a well fourteen feet deep, three by four feet outside measurement. The four posts should be four by four inches, and the planks two inches thick, fastened on with heavy spikes. Dig down until there is danger of caving, and then put in the curb, with planks enough on to reach the surface of the ground. Afterwards dig the earth from the inside of the curb, and put on the planks as fast as needed. In some soils that are loose, the weight of the curb will settle it down as the work progresses ; should it not, drive on the posts. Such a curb, made of sound hemlock, will last for years, and give pleasant water from the first.

SECURING THE WELL-BUCKET.

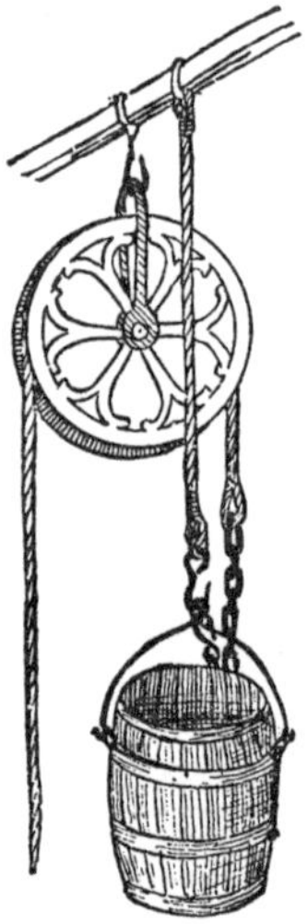
Fig. 124.

One who has much experience with well-buckets, will find they are often set down outside of the curb, and not always in a clean place. In this manner the water in the well may be fouled with clay, if with nothing worse. Every person should be very careful to avoid anything that may in any degree tend to impair the purity of the water in a well. One way to secure this end is to have the bucket always in a safe place. This may be done by fixing a cord or a chain to the beam over the pulley, or to the stirrup of the pulley, and fastening a hook to its lower end, upon which the bucket should always be hung when not in use. This arrangement for the well-bucket is made plain by figure 124.

CURB WITH A BUCKET SHELF.

Another device for keeping the bucket clean is shown in figure 125. An iron plate of suitable size is held on the end of an arm fastened at right angles to an upright iron rod. The bottom of this rod rests upon an iron projecting from the corner of the curb, and the top is held in place by an eye-rod. The filled bucket is raised high enough so that the plate is placed directly under it. Let up on the windlass when the bucket is secure on the

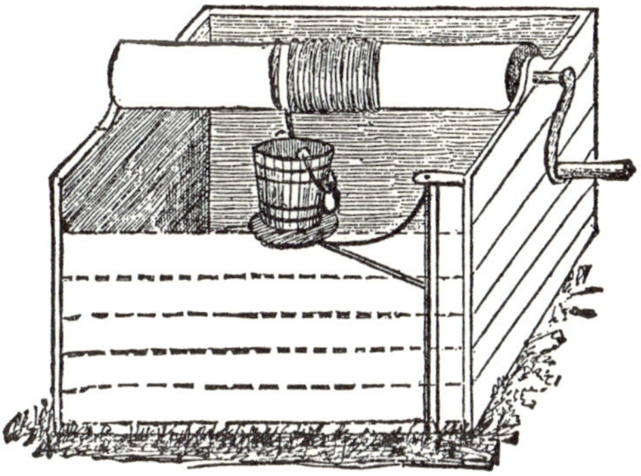

Fig. 125.—A BUCKET SHELF.

plate, and it may be swung to one side without straining the back, or danger of slipping when it is icy.

COVERED WELL-CURBS.

Figure 126 is a desirable covering for a well-curb. The upper part of the curb is floored over, except about a foot and a half in the center. Cleats, *r, r,* are nailed along two opposite sides of the bucket-holes and upon these, at

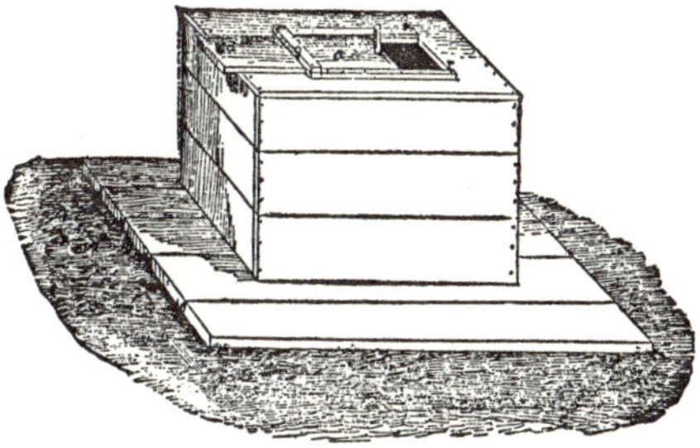

Fig. 126.—A WELL-CURB COVERING.

one end, is placed a strip, *g.* A wide board is made to fit and slide in the grooves formed by the cleats. The wooden pins, *a*, *a*, project above the cover, and answer as handles for sliding it as desired.

Figure 127 shows a covered well curb, which is safe against worms, frogs and other vermin; and also against the entrance of surface water, leaves and other objects. The wall of the well is carried up to the surface of the

Fig.—127.—A SAFE WELL CURB.

ground, and clean gravel is spread around it and beaten down firmly. A frame of four by four-inch chestnut is then bedded down level with this surface, and a floor of two-inch planks, with matched edges, is laid down, projecting a little over the raised gravel, as shown in the engraving. The ground slopes from the floor in all directions, and should be neatly sodded. The best covering, being indestructible, is a large flagstone; or, it may be made with several pieces and bedded in mortar. The curb is built around the well, large enough to give standing room for the bucket at one corner; it should be no higher than is convenient to reach over it to use the bucket. A spout is fixed to the front, into which the bucket is emptied without lifting it over the curb. The curb is protected on top with one fixed and one hinged wire gauze covered frame. The hinged one is thrown back and rests against a support, if desired, as shown, or it may fall entirely back upon the other one. This wire gauze admits air, but keeps out leaves and other trash, which is blown about by the winds. An open curb like this keeps the air in the well pure, the water clean, and in some respects it is preferable to a pump.

IMPURE WATER IN WELLS.

It becomes more and more evident each year that much of the sickness prevalent in the country is directly attributable to the quality of the water. By carefully studying the matter, it is found that in nine cases out of ten typhoid fevers originate in families whose water supply is from a well, into which impure water comes. This may be from the farm yard, and quite generally such is the case. For some years the water in a well near the house may be pure and wholesome, but by-and-bye the soil between it and the barn-yard will become so impregnated with pollution that an unhealthful quality will be imparted to it, and disease will result from its use. This is almost sure to be the case when the distance between the two is not great, because, as a general thing, the bottom of the well is lower than the yard, and the drainage from the latter will extend in all directions through the most porous strata of soil, and when it reaches the well, it will naturally flow into it as a reservoir. No matter how pure the water may have been when the well was first dug, sooner or later it will be contaminated by water flowing through the soil from barn-yards and cesspools located anywhere near it. A case is on record in which four children died from diphtheria. An examination by the physician proved that the slops from the kitchen had so filled the soil for a distance of twenty feet between the back door, out of which they were thrown, and the well, that the water in the latter was polluted by foul gases, and from the use of it diphtheria had certainly resulted. When making a well, have it, if possible, above the barn-yard, and let the drainage be from it rather than into it. Arrange a place for slops with a cement bottom and sides, from which glazed pipes, cemented together, allow the unhealthy matter to flow off and away from the well.

HOOK FOR CLEANING WELLS.

Every farmer who has open wells, knows how difficult and tiresome a task it is, to extricate articles which have fallen into them, but figure 128 shows a contrivance which has been used successfully. Find the depth of the well and cut off as many eight or ten-foot lengths, four inches wide, of inch boards, as will, when fastened together, reach to the bottom of the well. Sharpen the end of one length to a point, as *a* in the engraving; bolt or nail a cross-piece *b*, three or four inches above the point *a*, making one side a little longer than the other, and about three inches shorter than the radius of the well. Fasten upon the cross-piece three pieces of chain, each about ten inches long, at equal distance from each other. Make double hooks out of one-eighth inch wire, or old bucket bails; sharpen the ends, and attach them to the chains. Bolt the lengths of boards together closely; let the cross-piece and hooks down into the well, tightening the bolt at the end of each length as it passes, until the point reaches the bottom. Now turn the contrivance, causing the hook to describe several circles at the bottom of the well. The article sought for will probably be caught by the hooks;

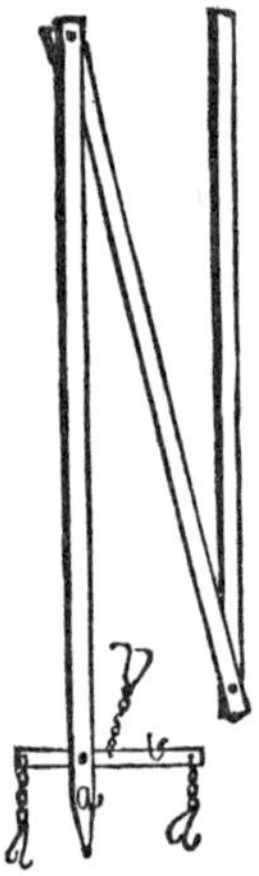

Fig. 128.

besides a good many other things not looked for will perhaps be brought up.

A NON-FREEZING PUMP.

One of the simplest methods of preventing a pump from freezing is shown in figure 129. The pump is boxed from the platform to six inches or more above the spout, the box being made large enough to admit of a

Fig. 129.—PUMP PROTECTOR.

packing of sawdust or spent tan bark between it and the pump-stock; or the pump-stock can be well wrapped with heavy hardware paper and then boxed tightly, which will effectually keep out almost any ordinary degree of cold. It is well to have the platform double-boarded, running each layer of boards in opposite directions, and mounding up well around the platform with earth, to still further protect against cold.

AGITATION OF AIR IN WELLS.

One great objection to the old style of log pump is the non-ventilation of the well. The platform is made as tight and close-fitting as possible, to prevent dirt, vermin etc., from getting into the water. By the use of a chain pump there is enough to agitate the air and water and to prevent stagnation in either. By means of a cheap, simple contrivance, shown in figure 130, all wells may

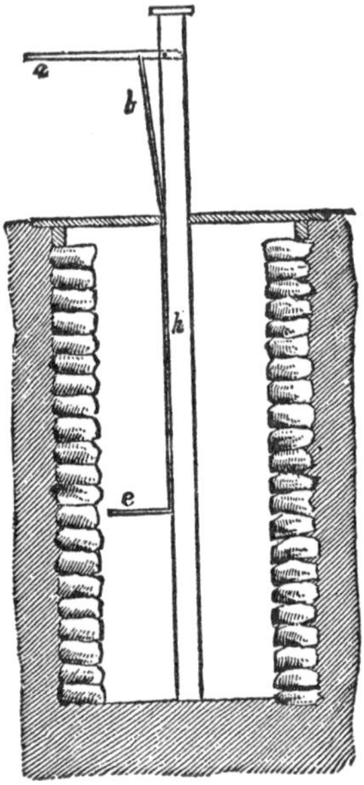

Fig. 130.—AGITATOR FOR WELL.

have an abundant supply of fresh air. In the illustration, *h* is the wooden or metal tube of a common lift or force pump; *a*, is the handle to which is attached, three or four inches from where it is hinged to the pump, a small wooden or metal rod, *b*. If this rod is of wood, it need not be over three-quarters of an inch in diameter, and if a metal one three-eights of an inch will answer. It runs from the handle downward to and along side of the pump-tube, shown at *h*, passing through two or more closely-fitting staples, and extends to within two feet of the high water mark. It is provided at the lower end with an arm, or more properly speaking, a fan, *e*, which should be of some light material, such as a thin board or piece of sheet iron or tin, eight or nine inches square. It is evident that the act of pumping will move this fan up and down, from three to five inches at each stroke of the handle, producing a movement of the air within the well. By continuing the rod downward for a few feet, and attaching to the end a block of wood two or three inches square, the water will also be sufficiently agitated to prevent stagnation.

DEEPENING WELLS.

Many wells which fail during long drouths, could be made, by deepening a few feet, to yield an abundant and unfailing supply of water. But it is difficult to accomplish this by ordinary means, without endangering the wall with which the well is lined. Figures 131 to 135 show a set of appliances by which the work may be safely done without danger to the wall, even in sandy or gravelly soil. Figure 131 is a sort of well-auger of galvanized iron, five inches in diameter, and of any desired length, from fourteen to twenty inches. Before it is bent in shape, a bias strip is cut from its lower edge, giving it the shape shown in the engraving. The rod by which it is worked is of wrought iron pipe one inch in diameter. A T is screwed on its summit, to receive the handle, of ash, or other teugh wood. Figure 132 is a cylinder, also of galvanized sheet-iron, six inches in diameter and two feet long. It is reinforced at each end by iron bands riveted on, and is perforated throughout with thin slits for the admission of water when in position. Figure 133 is the head of the auger. It is of inch board, upon which is screwed a flange with a thread, to receive the lower end of the hollow rod. Figuro 135 represents a cross-section of this head-piece. At the lower end of the auger-tube is the piece shown in figure 134. This is a circular piece of galvanized iron, cut five inches in diameter, slitted from one side to the center, and the cut edges bent to spiral or screw-shape. This is soldered

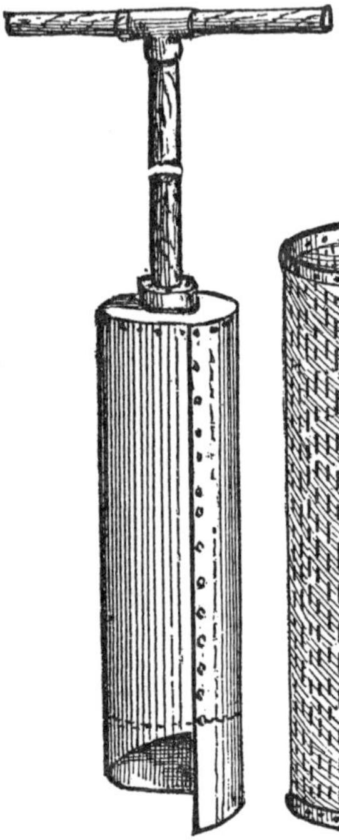

Fig. 131.—WELL AUGER.

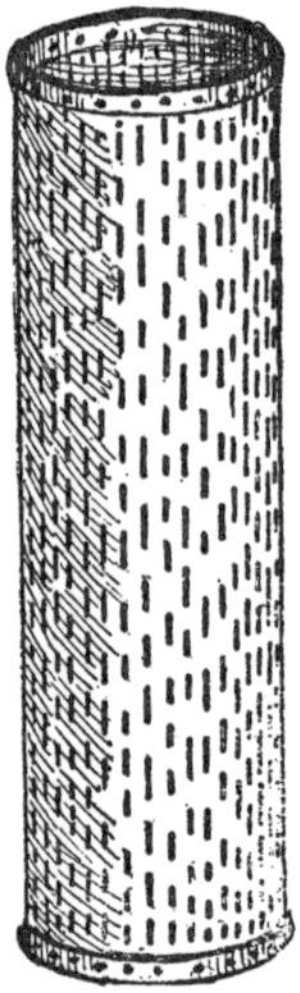

Fig. 132.—CYLINDER.

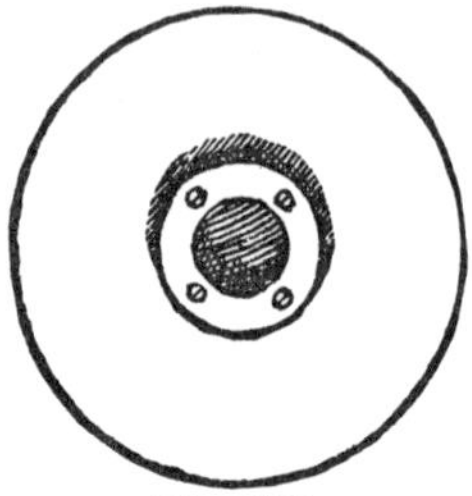

Fig. 133.

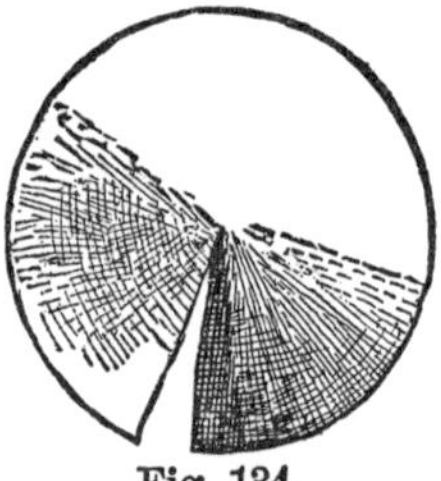

Fig. 134.

Fig. 135.

into the lower end of the auger-tube, as shown by dotted lines in figure 131. A large hole on one side near the top, not shown in the engraving, serves to empty the tube of sand and dirt. To operate this, the cylinder is first pushed down as far as practicable into the bottom of the well. With the auger the earth is removed from inside the cylinder. As the work of excavation proceeds, the cylinder is pushed down until its upper edge is level with the bottom of the well. If a sufficient vein of water is not then reached, the boring goes on, and a second cylinder follows the first. This makes an additional depth of four feet, which is generally sufficient. If not, the process can be continued by providing additional cylinders, and splicing the auger-stem until a permanent water-vein is found.

DIGGING A WELL.

A hole is dug down and the earth thrown out as far as could be done, and then a ladder is rigged up on three stakes as shown in figure 136. A pulley is attached to one round, a cord thrown over it and fastened

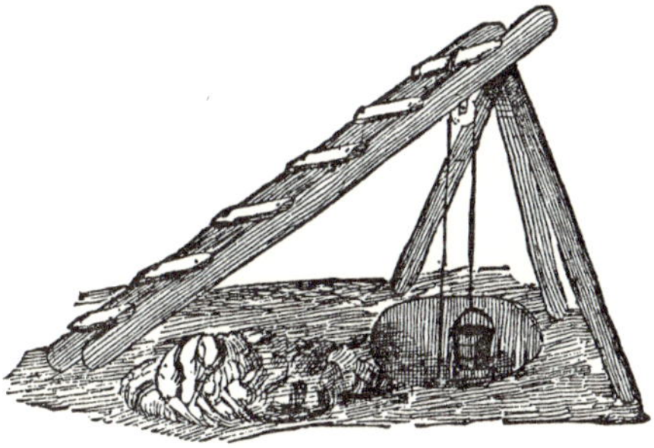

Fig. 136.—DIGGING A DAKOTA WELL.

to a pail, the other end of the rope reaching into the well. The pail is filled and drawn to the surface, where it is swung to one side, emptied and returned for another load. The upper end of the ladder should be elevated about six feet above the ground.

HOW TO BUILD A CISTERN.

Every part around the surface of a cistern should be made close. The beams which support the floor should be bedded in the wall, or shoulder of the cistern, and covered with lime or cement mortar, leaving a smooth surface all around the first floor. This should then be covered with a second floor, raised eight or ten inches on a frame of two by ten joists, made of cedar or chestnut.

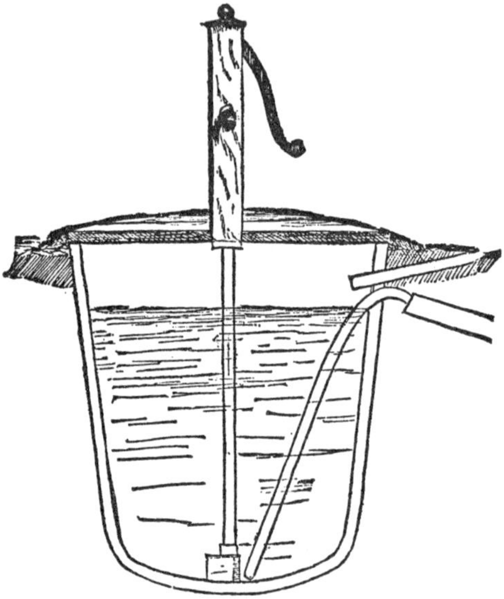

Fig. 137.—FROST-PROOF CISTERN.

The earth should be packed closely against this frame, and the top floor should extend a few inches beyond the frame all around. The cistern is then frost and vermin proof. Another important point is to get rid of the sediment that gathers at the bottom of every cistern. This is done by carrying the overflow pipe to the bottom of the cistern on a line with the inlet pipe, and thus forming a current which disturbs the sediment and carries it into the overflow. This is shown in figure 137, also the arrangement of the draw-pipe, which should have a fine wire strainer on the end, and should rest upon a support near the bottom of a fine strainer, at least two feet high. A piece of one-quarter inch mesh of galvanized wire gauze, bent into a pipe a foot in diameter, and covered with thick flannel cloth, doubled, makes a filter for the water.

WATER IN THE BARN YARD.

Water in the barn yard is a great economy and convenience; by a little management it can be secured with ease. The difficulties in the way, are chiefly in bringing the water down hill, over an elevation midway, and

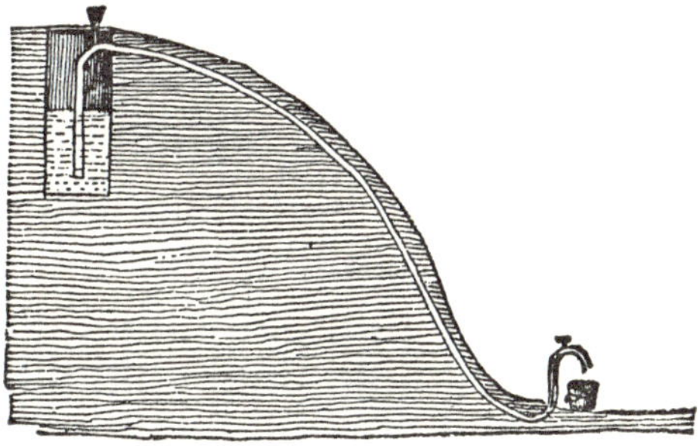

Fig. 138.—CONVEYING WATER BY SIPHON.

in bringing it up hill, from a spring or well below the level. In the former case, a siphon must be used, as shown in figure 138. It consists of a pipe, one end of which is under the surface of the water in a well or spring, and carried over a ridge, and down a slope, to a lower level. This method is open to one objection, which nearly always gives trouble. There is more or less air dissolved in water; this dissolved air escapes, as the water is flowing through the pipe, and gathers at the highest point, where it stops the flow. To remove this air, the following arrangement is made. A short pipe, furnished with a stop cock and a funnel, is fitted to the highest point of the siphon. When the flow begins to be obstructed by air, the stop cock at the lower end of the pipe is shut, and that at the top is opened.

The pipe is then filled with water through the funnel. The water is held in the pipe by a check valve at the bottom of the well pipe. The air is thus driven out of the siphon, and the top stop-cock is shut. The flow is started by opening the lower stop-cock, and all goes

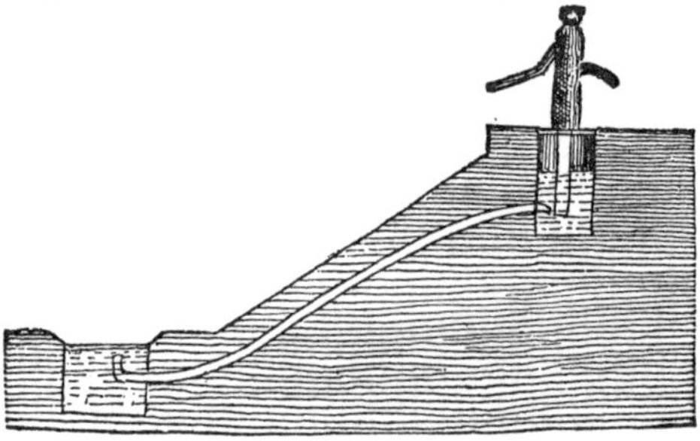

Fig. 139.—RAISING WATER BY FORCE PUMP.

on again, until the air gathers in the pipe once more, when the remedy is repeated. To draw water up hill, by a pump, the method seen in figure 139 is used : The pipe, having a check valve at the bottom, is laid from the spring, up the incline, and connected with a force pump, in a dry well, at the top. Water can be raised in this manner, from about twenty-eight or thirty feet below the bottom of the dry well, and for a distance of two hundred and fifty, or three hundred feet, or more. The linear distance is not an obstacle, except for the friction in the pipes ; it is the perpendicular height alone, which gives serious trouble, and about twenty-eight or thirty feet, is all that can be overcome by means of a suction pump. A force pump is useful to raise the water eight or ten feet more than this, when necessary. This method is shown in the engraving.

WOODEN WATER PIPES.

For conveying water any distances less than fifteen rods, and where the amount desired is greater than can be supplied by a half-inch pipe, wooden tubing will be found cheaper than iron, lead, or other metallic pipes. Wooden tubing, of from one and a quarter to two-inch bore, may be obtained of all hardware dealers. In purchasing observe that the ends are iron-banded, to prevent splitting when placed together, and to prevent the tubes from bursting when under a heavy head of water. Before the pipe is laid, it is best to give it one or two coats of oil; even crude petroleum will do; this adds greatly to the durability. In pipes through which there is a constant flow of water, there is little danger of decay; in fact, some old-fashioned pump logs which have been removed after nearly fifty years of use, were found sound on the inside. Wooden, as well as other pipes conveying water, should be laid below the frost line. If the water is intended for drinking purposes, place the pipe at least three feet under ground, and if in sandy, porous soils, to a still greater depth. After the pipe is in position, and before the water is admitted, pour hot coal-tar over it, especially at each joint, which is readily done by using a watering pot or an old tea or coffee-pot. Always test wooden and other pipes after they are laid, by admitting water before covering them with soil, in order that a leak, if found, may be easily stopped.

FILTERS FOR FAMILY USE.

Almost every country store is in more or less direct communication with some pottery, where salt-glazed ware is made. Lead-glazed ware should be avoided, but the salt-glazed is both cheap and safe. Any pottery will furnish to order, or they may have them on hand, five or six gallon cylindrical jars of glazed ware, having a spigot hole in the side close to the bottom, and the usual jar lid. A common flower pot of large size should be selected.

which will just fit in the top of the jar, as shown in figure 140. This pot is the filter, and it is thus arranged: The bottom is covered by a circular piece of

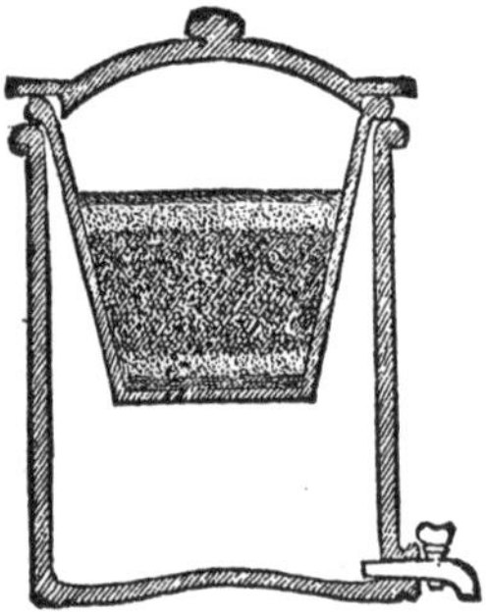

Fig. 140.
FARM WATER FILTER.

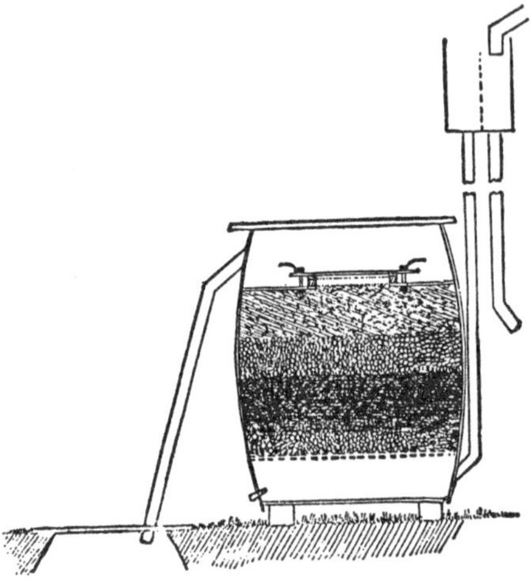

Fig. 141.
A SERVICEABLE FILTER.

thick woolen felt, or two or three pieces of blanket, upon this is placed a layer an inch thick of well-washed sand. Note that the sand, being well washed, the felt or blanket pieces should be so also. Now upon the sand, freshly burned, soft wood charcoal, which should be freshly heated, free from all dust, and about the size of grains of wheat, should be laid in to the depth of six inches more, and upon this an inch of sand, and another pad of felt or blanket to top off with. Fit up two flower pots for each filter, and keep those not in use covered and clean. The water is poured into the flower pot. A clean lump of clear ice, whole or broken up, may be placed in the jar below the pot, and then the water is fit for anybody's use.

The filter shown in figure 141 is made of a stout oak barrel with iron hoops. The head is taken out carefully and a number of holes are bored through it with a half-inch auger, then five or six oaken blocks, about three inches long, are nailed to the under side. It is then placed in the barrel for a false bottom. On this spread a layer of coarse gravel about two inches thick, then another of finer gravel; on this spread eight inches of charcoal; then add a six-inch layer of gravel, and on top place washed sand up to within an inch of the overflow pipe. Over this sand fit in the barrel a cover made of inch pine boards. In the center of this cut an opening ten inches square. Then make a low frame a little larger than the opening, cover both sides with cheese cloth, and fasten securely over the open space in the head, but in such a manner that it can be easily removed again. The object of this covering being to prevent sand from escaping into the cistern, it becomes sometimes necessary, after heavy rains, to take up the frame and wash the cloth.

The rain water flows into the barrel through a pipe between the real and the false bottom. About four feet above the cask the leader from the roof should enter a tin box, with a partition in the middle that comes within about two inches of the top. This partition separates the pipe that flows into the cask from the waste pipe, and the leader from the roof can be made to discharge on either side, as may be desired. Near the bottom of the barrel should be a large faucet or bung-hole, through which all the water may be drawn off and the filter cleaned. By pulling out the bung or opening the faucet, and, after the water has run out, pouring several bucketfuls of water on the sand at the top, all impurities are washed out and carried off; in fact, it is best to let out the water after every rain. If this filter is well made, and the cask painted, it will last many years and do good service.

CONNECTING CISTERNS.

When it is desired to connect a new cistern with an old one without loss of water, it can be done as shown in figure 142. Whatever the distance apart, provide a two-

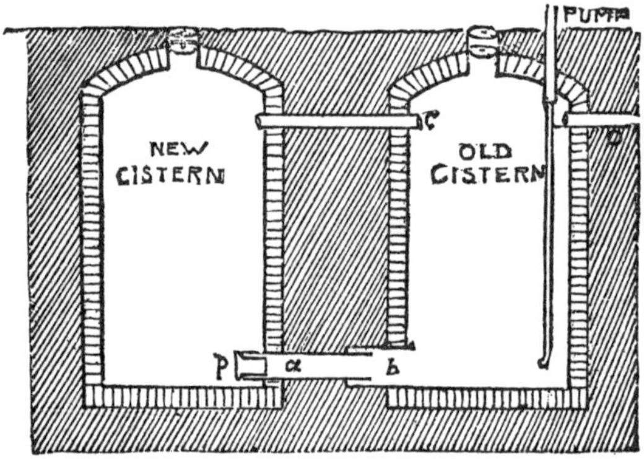

Fig. 142.—CONNECTING TWO CISTERNS.

inch iron pipe, *a*, long enough to extend from the inside of the new well to the outside of the old one, and fit upon the right end of it by screw thread the pipe, *b*, long enough to extend well through the old wall. Build *a* into the new well, and close its left end with a wooden plug, *p*. When ready, pump or syphon the water from the old to the new cistern. Then open the old wall, screw *b* on to *a*, and cement around *b*. When ready, with a rod or bar, knock out the plug, which will float to the top, and the water will stand at a level in both cisterns. A connecting five or six-inch glaze ware or iron pipe, *c*, should be put in on a level with the overflow pipe, *o*. One pump and one overflow pipe answer for both cisterns.

BUILD AND DIMENSION OF CISTERNS.

In a stiff clay soil a small cistern of twenty to forty barrels capacity might be safely cemented directly to the earth, but in ordinary soils and for larger cisterns, a good four-inch-wall of hard brick is on the whole the cheapest. It is important to make the excavation smooth, so that the bricks can be pressed firmly against the earth; otherwise these will be pushed out and the cement cracked, causing a leak. As to the dimensions, a cistern should be about one-fourth deeper below the spring of the arch, than its width inside. By this rule a cistern eight feet wide will be ten feet deep below the arch. At the top is a cast iron ring, twenty inches in diameter, for the man-hole, covered with a tight fitting cast iron lid. The ring has a flange two inches wide extending out over the brick. The capacity of a cistern needed to save all the water from a given extent of roof, will depend on the total annual rainfall, its distribution throughout the year, and the regularity with which it is used. A roof ninety feet by twenty feet contains eighteen hundred square feet. This is supposed to be the measure of the building on the ground and not the shingled surface. In the vicinity of New York the average annual rainfall is about forty-two inches, or three and a half feet. This would give sixty-three hundred cubic feet of water (1,800 ft. $\times$ $3^1/_2$=6,300). Since in that climate the rain is distributed pretty regularly through the year, it would only be necessary to provide storage capacity for about one-third of the rainfall of the year, or twenty-one hundred cubic feet. This divided by four and one-fifth (the approximate number of cubic feet in a barrel of thirty-one and a half gallons) gives five hundred barrels, and this quantity of water demands a cistern, thirteen feet diameter, to be nearly sixteen feet deep below the arch, or a square one, thirteen feet across, to be nearly twelve and a half feet deep ; or a round one, fifteen feet in diameter, would need to be about twelve feet deep. In the far West—in fact, in most places west of the Missouri—the rainfall is largely during the six months beginning with March, and cisterns need a greater storage capacity.

CISTERNS WITH FILTERS.

Complaints are frequent of the impure water of cisterns. This is inevitable under the careless management of these useful additions to the water supply, and is a fruitful source of what are called "malarial dis-

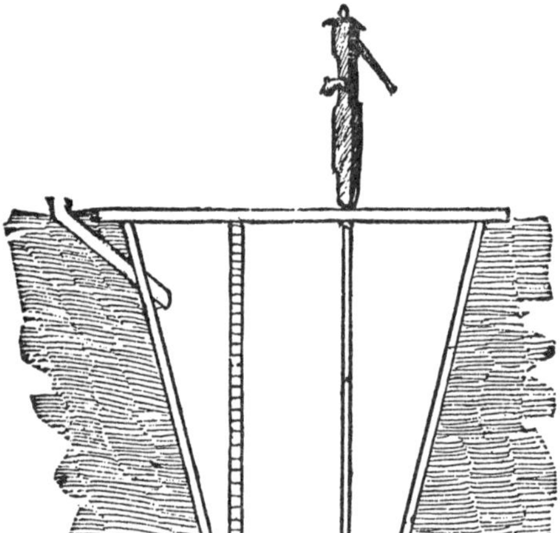

Fig. 143.—A COMPLETE CISTERN.

eases." A roof gathers a large quantity of impure matter, dead insects, droppings of birds, dust, dead leaves, pollen from trees, etc., etc., all of which are washed into the cistern, unless some means are taken to prevent it. Even then the water should be filtered before it is used for culinary purposes. One way of preventing foul matter from entering the cistern, is to have the leader movable, and swing from a waste pipe to the cistern pipe, shown on the left side of figure 143. In dry weather the pipe

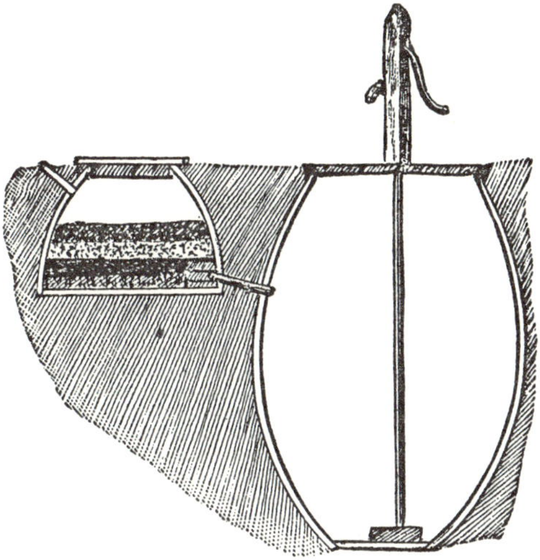

Fig. 144.—COMPLETE CISTERN AND FILTER.

is turned over the waste, and after the rain has fallen for a sufficient time to wash off the roofs and gutters, it is turned into the cistern pipe. The cistern, figure 143 is provided with a soft brick wall laid in cement, through which the water filters, coming out by the pump perfectly pure, and free from unpleasant odors. Rain water

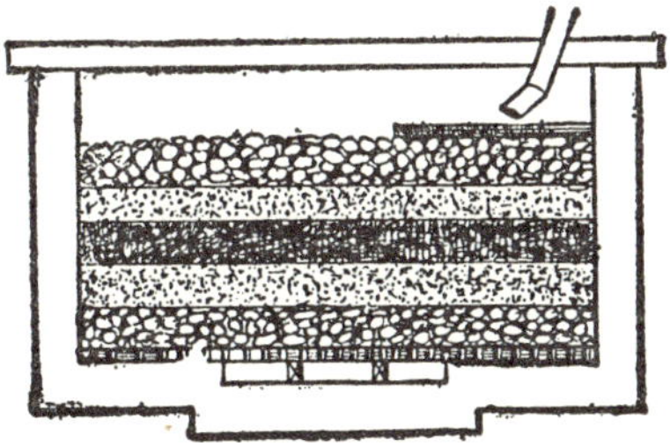

Fig. 145.—FILTER FOR A BARN CISTERN.

standing for months in impurities and filth, cannot always be purified by simply soaking through a brick wall, but should be filtered as soon as it falls. The main cistern, figure 144, is made egg-shaped, to hold one hundred barrels. The filter is flat-bottomed. The end of the pipe from the filter to the cistern is built solid

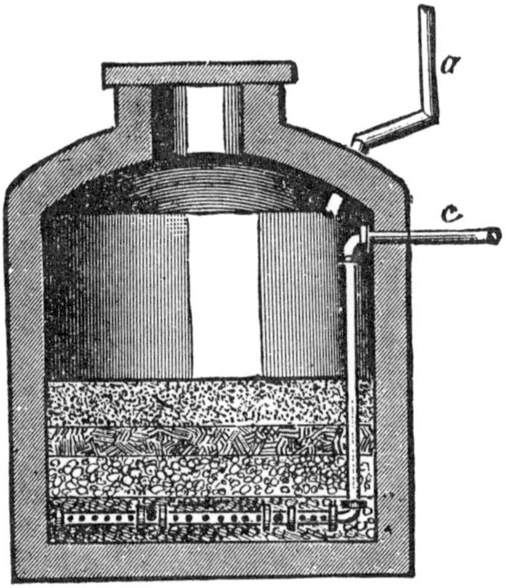

Fig. 146.—A HOUSE FILTERING CISTERN.

around the end with brick. All water has to pass through the brick. The filter is filled half full with charcoal, sand and gravel in layers—one layer of each—the charcoal covering the bricks, then sand and gravel on top. The water, as soon as it falls, begins to filter and passes into the cistern, where it stands free from impurities. The filter is built to hold twenty-five barrels of water, but is half full of the filtering material.

Figure 145 shows a good filter for a barn cistern. The top of it consists of broken stones, with a flat stone to receive the influx, so placed as to prevent heavy rains from disturbing the broken stones. This has a cover, movable in part, to permit it to be cleaned out occasionally. Figure 146 is a filtering cistern for a house. The inlet pipe is at *a*, the draw pipe is at *c*. and this is connected with a set of cross-pipes, laid in the coarse gravel in the bottom, and pierced with a number of small holes,

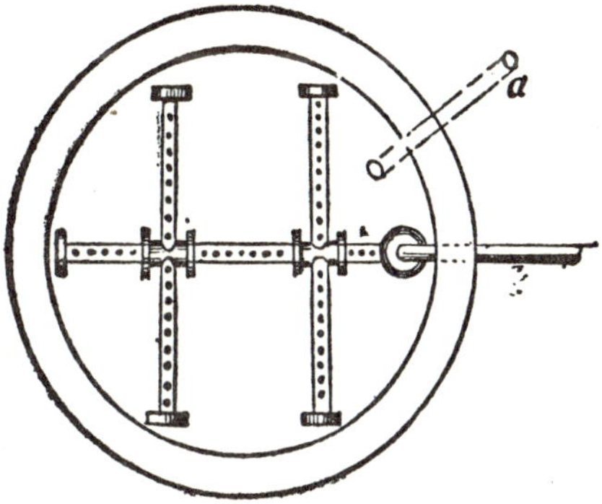

Fig. 147.—PIPES FOR HOUSE CISTERN.

as seen in figure 147, by which the outflow is made quite easy and abundant.

CHAPTER VI.

APPLIANCES FOR HANDLING HAY AND CORN FODDER.

REVOLVING HORSE RAKE.

Figure 148 shows a strong, cheap and efficient horse rake. It is especially useful in raking corn-stalks that have been cut by a mower or otherwise, and tall reeds and other rubbish, which it is desirous to rake into windrows preparatory to burning. It can also be adapted to the raking of hay and straw, by making the teeth lighter and placing them six inches or less apart.

Figure 149 represents the rake and shafts, *a* being a six by six-inch beam, ten feet long. This revolving rake can be made longer or shorter as desired, but when more than nine or ten feet long, it is not easily drawn through ordinary farm gates. The teeth are made of some kind of tough wood, well seasoned, two inches square, and the

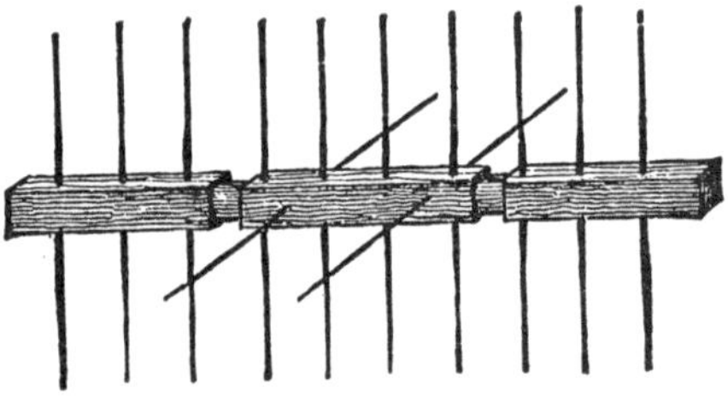

Fig. 148.—BODY OF RAKE.

pieces cut about four feet long. They are then tapered slightly toward the ends, and trimmed in the middle to fit in holes bored with a two-inch auger ; thus prepared, they are inserted one foot apart, and secured in place with light bolts. At *b, b,* the beam is rounded to form journals, and around these the ends of the shafts can be

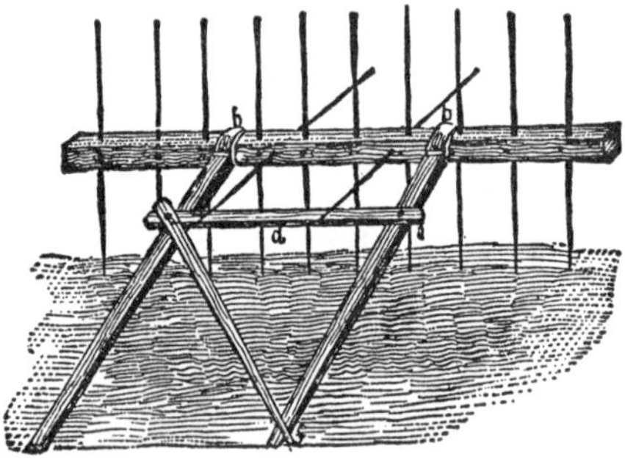

Fig. 149.—RAKE WITH SHAFTS.

bent, as seen in figure 149, or pieces of old iron, as the tire of an old wheel, may be curved round and secured to the shafts. Two stout pieces of the same length as the teeth, and at right angle to those, are inserted between the shafts. These rest on the lever, *d,* when the rake is moving, and serve to hold it in position with the teeth pointed toward the ground. The lever is hinged to a shaft at *e* by a bolt, and by pulling the handle, *f*, when in motion, the support is taken from the check teeth,

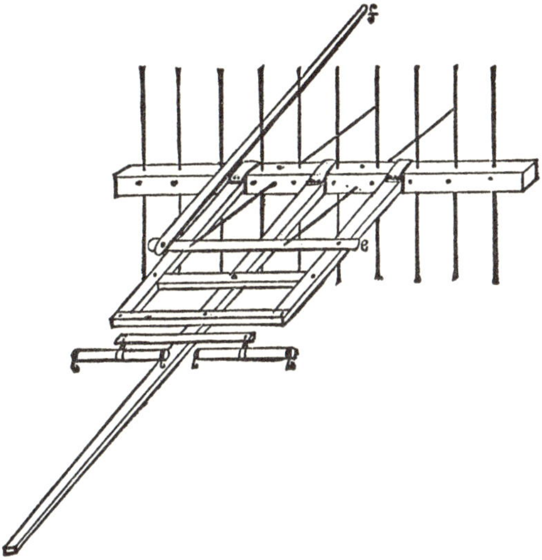

Fig. 150.—RAKE FOR TWO HORSES.

and the rake turns, depositing its load, and bringing the other row of teeth to the ground.

An implement of the above dimensions is too heavy for one horse; hence the shafts are intended to be hooked to the hind axle of a naked wagon, and thus worked by a team, the driver riding on the wagon and operating the lever whenever the rake is full.

The rake can also be made with a pole, so as to hitch a team directly to it, as shown in figure 150. The beam should then have three journals instead of two, and the number of teeth even, so that the pole can be attached at the middle.

CARE OF MOWING MACHINES.

Take up the wear of the boxes by removing the stuffing. If the journals have too much play they cut fast. But if the boxes fit too closely, they will heat. The thickness of newspaper all around each is sufficient play. Examine all the nuts and tighten any that are loose. A loose nut will cause the machine to wear or break; and will lose off in the field, causing a vexatious delay at the least. If any nut is very loose, place a leather washer under it, and sink the nut in. Make quite sure that there is no weak place in the whiffletrees. They always break at the wrong time, and may allow the machine and the team to mire down in a muddy spot. Mind the sickles. Every mower should have three sickles, so that as soon as one becomes dulled, another may be put in its place. A dull sickle does "ragged" cutting, and will increase the wear and draft of the machine one half. Observe if the sickle bar is not warped; and if the points of the sections are in a perfectly straight line. A section out of line will wear fast and increase the draft. Try the sections and tighten any that are loose. A loose section is apt to cause a breakage. A loose guard will produce the same result. See that the tool-box contains claw-hammer, pincers, file, sections, rivets, bolts, wire and nails; the lack of these will often require a trip from the field to the tool-house. If any journals have rusted, use coal oil, every few minutes, for the first half hour, driving slowly; and it is well to use coal oil on the track of the sickle, to clear off the gum that gathers from the grass. The machine oil often sold is poor stuff; it is frequently necessary to add castor oil to give it body. If too much of the latter is used, however, it will gum. There is nothing more satisfactory than lard (unsalted) with castor oil added to give it a little body. If the lard is taken to the field hot, in the morning, the sun will keep it liquified during the day. It is not economy to be sparing in the use of oil; it should be applied quite often, and but little at a time. When much of it is applied at once, it runs from the journals, and holds dust, increasing instead of diminishing the wear. When stopping at noon, throw some grass over the sickle and the driver journals, if you cannot drive the machine into the shade. Do not mow too close. It dulls and wears the sickle, and gains nothing—what is gained in hay is more than lost in the aftermath. Drive slowly, but steadily, and thus get the most done with least wear of team and machine. Driving "in spurts" for half a day will wear the machine more than steady driving for two days. Keep the edges of the grass straight; in other words, cut the full width of the sickle, for otherwise you cannot do economical work. Using a mower properly lengthens its life and increases the amount of work it will do in a day.

SWEEP FOR GATHERING HAY.

The implement shown in figure 151 is made by having two by four inch pieces of twelve feet long for teeth,

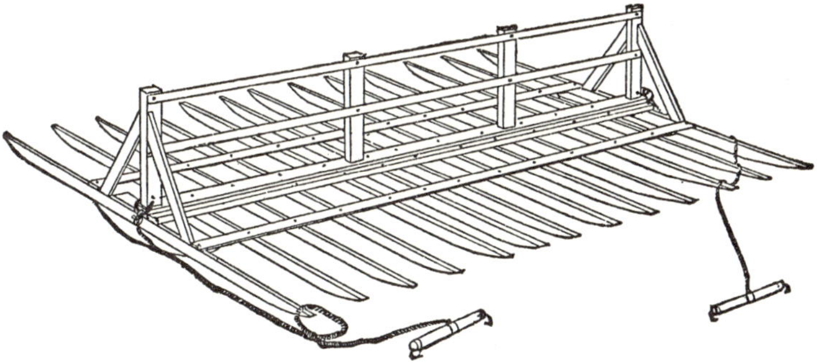

Fig. 151.—HAY SWEEP.

held together by a piece of two by six inch stuff, sixteen feet long, with a bolt through each tooth; two fence strips of inch stuff, six inches wide, keep them from spreading. Three or four upright two by four posts, four feet high, with cross fence strips, are set on the main beam. This holds the hay, and is braced at each end, as shown in the engraving. There is on the bottom of each end runner, a shoe one foot wide, two inches thick, and two or three feet long, to give the teeth a downward inclination. The teeth are tapered from the underside at each end, so as not to run into the ground. There is an iron ring at the bottom of each end post, to which ropes are fastened. These ropes are sixteen feet long or more, and a whiffletree is attached to the end of each. This contrivance takes up the hay to the stack, and picks up any dropped hay going back.

HAULING HAY OR STALKS.

Figure 152 shows a device for hauling an entire cock of hay. It is made thus: First, get a pole, elm if possible, ten or eleven feet long, and about four inches through at the butt. Peel off the bark, trim smooth,

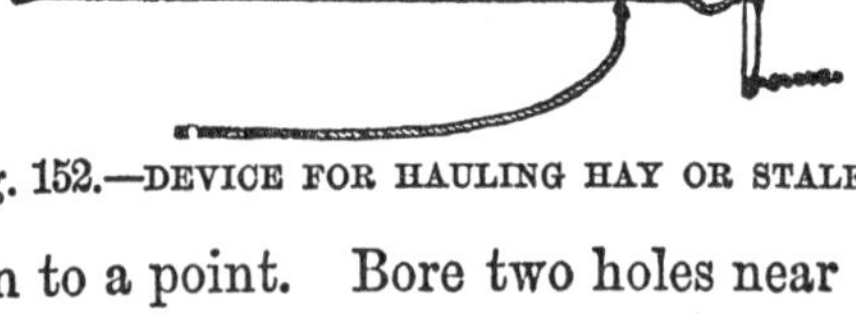

Fig. 152.—DEVICE FOR HAULING HAY OR STALKS.

and sharpen to a point. Bore two holes near each other at the butt; pass a short piece of rope through the pole, and tie to the link on a single tree. Bore another hole a foot from the end, and pass through it a long one-inch rope, shorter on one side, and tie a knot on the rope on each side of the pole. When ready to commence hauling push the pole under the hay-cock, then take the long end of the rope, and pass it along side the hay-cock, and under the point of the pole, then through a loop in the short end, and draw tight and tie. By this method, no hay is lost on the way; it cannot roll over, nor get tangled. There is no waste, no time is lost, and the hay is laid at the feet of the pitcher just as it stood in the field. This device may also be used for hauling corn fodder or unhusked stooks.

DERRICK FOR STACKING.

Figure 153 shows a derrick, which is very convenient in stacking hay out-doors. The two side-pieces are mor-

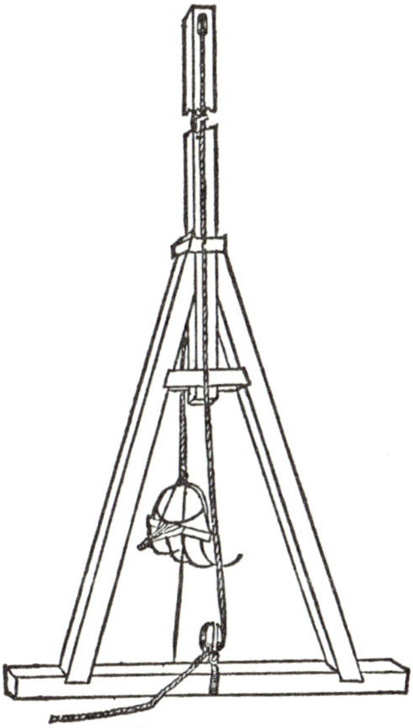

Fig. 153.—HAY DERRICK.

tised into the sill; the mast, or top stick, is not fastened, hence it can be taken out when moving any great distance. When moving only a few feet, hitch a horse to the sill, and drag it endways without taking it down, to where another stack is to be. The derrick and horse are on one side of the stack, and the load of hay on the other. The derrick stands at almost forty-five degrees, and is held in place by guy ropes—two opposite to the load, and one on the same side as the load. A solid piece of plank is used for the foot of the mast, which is mortised into it. It is best to put it together with bolts.

HAY CARRIER FOR HORSE FORK.

Figure 154 shows an ingenious device for returning a horse hay-fork from the hay-mow to the loaded wagon. It consists of a wire rope, *C*, stretched from the end of the track, *A*, to a wooden cylinder, *B*, four inches in

Fig. 154.—IMPROVED HAY-CARRIER.

diameter and sixteen inches long, around which a few turns are given. Two short stakes, *D*, *D*, cut from a four-by-four-inch scantling and driven slantingly into the ground, hold the roller in position. A grooved pulley, *E*, runs freely on the wire, and from its axis is suspended a fifty pound weight, *F*. The rope, *G*. runs over the pulley, *H*, which is firmly attached to the lower side of the track. The wire-rope is made of three wire clothes-lines twisted together. When in use, the upper end of the cord is attached to the rope which carries the fork. It is thus carried up with the loaded fork, and brings it back by gravitation when empty.

HAY BARRACKS.

Figure 155 shows barracks constructed by setting four posts, of chestnut, or white oak, twenty or twenty-five feet long, straight, partly squared to eight inches through, either three feet in the ground or upon sills. If upon sills, these are hewed upon one side and at the

Fig. 155.—BARRACK WITH BOARD ROOF.

ends, where they are halved together. In doing this, it is well to pin the ends with two inch oak tree-nails, which should stand up three or four inches above the sills when in place. Then when the posts are set at the corners, the pins will enter holes bored in the center of each post, and hold them in position. They will last as long as the posts and sills. For a temporary purpose, the posts may be simply set in the ground, twelve feet apart; but if permanency is desired, it is best to use sills, set level upon a flat stone at each corner, and supported in the middle. The posts must, moreover, be braced to the sills, either by diagonal braces, or straight rails, roughly squared, two by four, mortised into the posts at a height of four feet above the sills, or at a height of six feet, in case the barrack may be intended to be boarded up to make a stable. They make very good

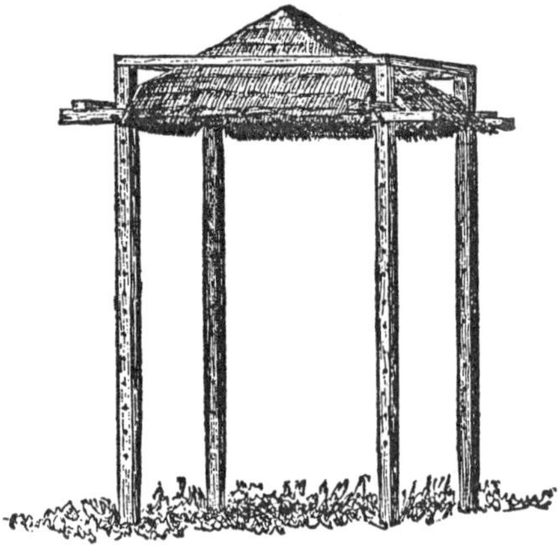

Fig. 156.
BARRACK WITH THATCHED ROOF.

Fig. 157.
MANNER OF RAISING.

shelter for young cattle or horses, the inside between the sills being filled up level with stones, and finished with a layer of cement concrete. The cover, or roof, must be as light as is consistent with strength and efficacy, and may be of boards, or thatch, the latter being by far the most picturesque, and being lighter, it is easier to raise and lower. Before the posts are set, they are bored with inch holes, either twelve or fourteen inches apart, from the top down, exactly in the middle line, each set perfectly level, and pass through the posts in the same direction. Four pins of three-quarter inch iron, fourteen inches long, turned up a little at one end, and bent slightly downward to prevent rolling, are the roof supports. They are put into a set of low holes on the outsides of the posts, and two straight oak rails, sixteen to twenty inches longer than the space between the posts, are laid upon them. Then across the ends of these, and outside the posts, two similar rails are laid, the ends being temporarily bound together at the corners. These form the plates for the roof. One-third pitch is usually given, and the ends extend ten inches, or a foot, beyond the plates. A good coat of paint will make the roof quite durable, and prevent the boards from warping. To make a thatched roof, figure 156, nice, straight, light hoop poles are selected, which, if too heavy, must be split. These are for rafters. If binding poles are used, they must be mere rods, like light whip stocks. The rafter poles are laid up and bound at the ends, and to the cross-poles with tarred rope-yarn, but nailed to the plates. They are placed about eighteen inches apart, but the light split cross poles, about a foot to fourteen inches apart. The straw is laid on in handfuls, beginning at the eaves, and bound with rope yarn to the cross poles, or in courses, and bound down by tying the tough, slender maple rods, to the cross poles. Of course, the straw is kept even, and in courses, butts outward, and trimmed evenly with shears. When laid, the straw must be well evened at the butts, and dampened so as to pack nicely and not break in handling. These covers should be as light as possible, and be consistent with strength. They are raised and lowered one corner at a time, which may usually be done by one man, though more conveniently by two. To raise the roof, a ladder of suitable length is set under the lower plate pole of one corner, as shown in figure 157, the end of the pole being allowed to pass through between the rounds of the ladder, which is then lifted either by main strength, or by a rail used as a lever, and held in position until some one going up another ladder, can lift the pin which supports it. This is, of course, done at each corner, and thus the roof is raised, one peg at a time. It is lowered in the same manner by reversing the operation.

SUPPORTS FOR STACKS.

In stacking straw or hay, when stock is permitted to feed upon it during the winter, it is unsafe to leave the stack without support. The danger is that the stacks may be undermined, and fall over upon the animals.

Fig. 158.—FRAME FOR STRAW OR HAY STACK.

This will not happen if a stout support is made, as shown in figure 158. A few strong posts are set firmly in the ground, and planks spiked on the side as shown; the cattle can eat the straw from between the planks, and may eat the stack entirely through without danger of its being buried by over-turning. When the crib thus made is filled, the stack is topped off in the usual manner, being well spread over the eaves to shed the rain, and, as it is eaten out below, the straw settles down gradually. It is quite easy to cover a stack so made with a roof, so as to form a very cheap barrack. In the summer, by a little change, this will make a good calf or sheep pen.

HOME-MADE HAY PRESS.

The press shown in figures 159, 160, and 161 may be made wholly of wood, hewn to the right size, and put together with wooden pins. The frame, figure 159, is four feet long inside of the posts, and three feet wide. The

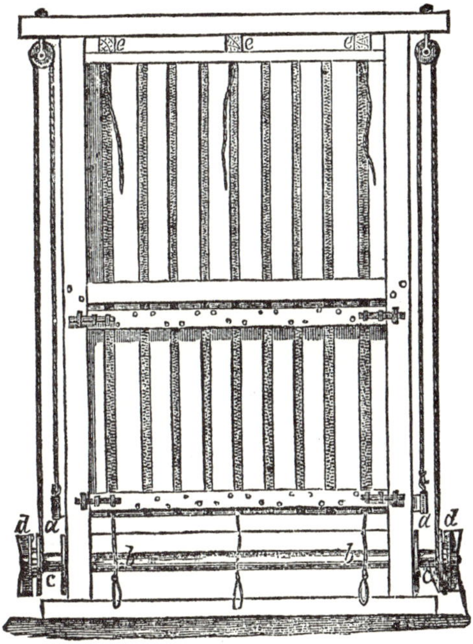

Fig. 159.—FRAME OF HAY PRESS.

height is eight feet. The movable bottom is raised by ropes which pass over pulleys or rollers, if no iron is to be used, and are wound upon the rollers at the bottom. This roller is moved by bars to be inserted in mortises cut in the roller, similar to the manner used in moving a windlass, or a capstan on shipboard. A movable door is made to fit the bottom of the press on one side, for tho purpose of removing the bale after it is pressed. The bale is bound with a strong cord, pieces of which are placed on the bottom and others on the top, as shown in

Fig. 160.—END VIEW OF HAY PRESS.

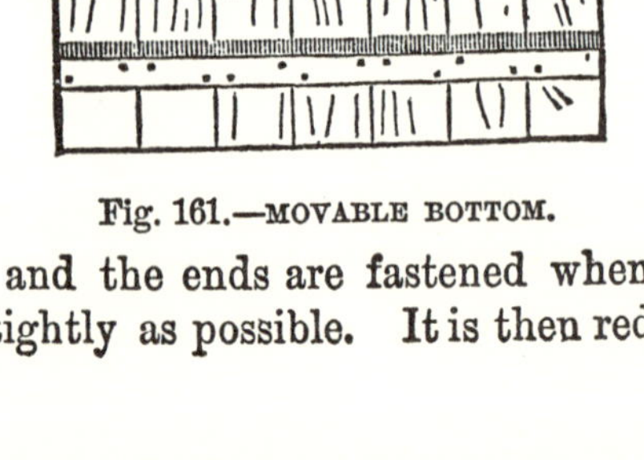

Fig. 161.—MOVABLE BOTTOM.

figure 159, and the ends are fastened when the bale is pressed as tightly as possible. It is then reduced to two and one-half feet in thickness, and eight of these bales will make a ton. The hay is easily transported in wagons when baled, and the press can be moved from one meadow to another as the hay is cut and pressed, or it will

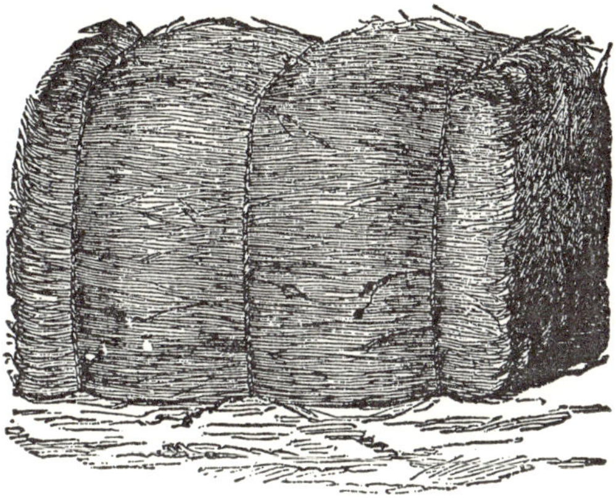

Fig. 162.—HAY BALE.

be more convenient at times to drive the cattle to the hay rather than move the hay to the cattle. Figure 160 shows the end view of the press, figure 161, the movable bottom, and figure 162, the pressed bale. When the iron can be procured without great expense, it might be well to use the pulleys and slotted wheels as here shown, but otherwise these parts may be made of wood.

TWISTING HAY AND STRAW.

The machine figures 163 to 167, consists of two two-by four bars, nine feet long, figure 163, straight and true, and of even thickness and width. They are bolted together at each end, and separated by a block four inches square and two inches thick, at one end *a*, and a piece of two-by-four stuff, three feet long, at *b*. A pulley is set at *c*, about thirty-nine inches from the end *a*. Figure 164 shows frame of back end (*A*, figure 167), *a* being a piece of two-by-four, four feet long; *b*, *b*, two uprights, one-by-four, six feet long; *c*, a two-by-four, three feet long; *d*, a one-by-four, three feet long; they are firmly

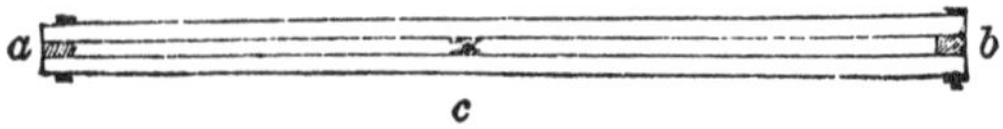

Fig. 163.—THE BARS, OR WAYS.

nailed together as shown, the upper edge of *c*, being half way up from bottom. Figure 165 is the same as figure 164, except the lower piece *a* is only three feet long. The five converging pieces are of some springy wood. They are attached by screws, three to the upper cross-pieces and two to the middle one, and prevent the hay going too fast out of the rack. Figure 166 shows a "follower" (*f*, figure 164), *a* being two-by-four, twelve inches long, *b*, two-by-four, twenty-six inches long, framed or halved on *a*; *c* is a brace of one-inch board; *d*, *d*, two pieces of board, the lower one eight inches wide, six inches long,

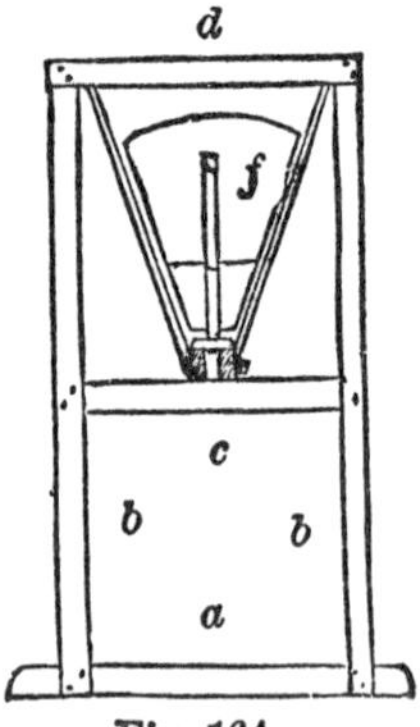

Fig. 164.

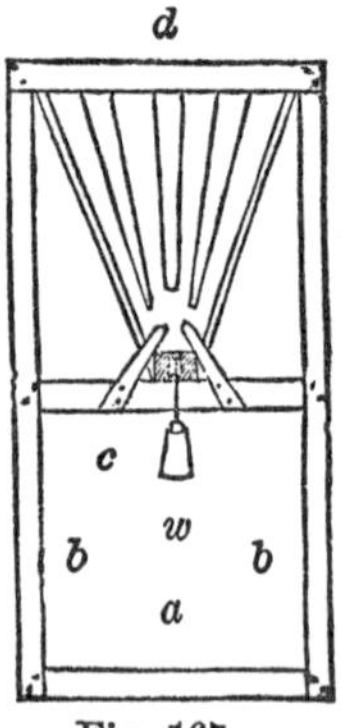

Fig. 165.

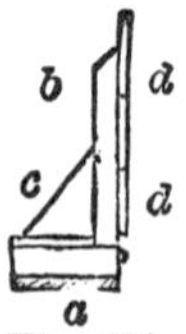

Fig. 166.

the upper twelve inches wide, sixteen inches long; they are firmly attached to *b* by nails and screws as shown, and the brace *c* is then nailed in place. The "twister" is made of seven-sixteenths round iron, nine inches long from crank to hook, five inch crank, three and a half inch handle. The hook is turned so as to have a twist like a corkscrew, so that it will work in and catch the hay up of itself; two washers are put on that fit the rod snugly and four inches apart, by placing shaft in a vice; a nick with a cold chisel on each side of shaft on outside of each washer, will keep them in place. This is made fast upon a piece of two-by-four, twelve inches long, the upper end grooved out so that the shaft will set in about half way, then beveled off as shown; the shaft is fastened in place by a couple of strips of hoop

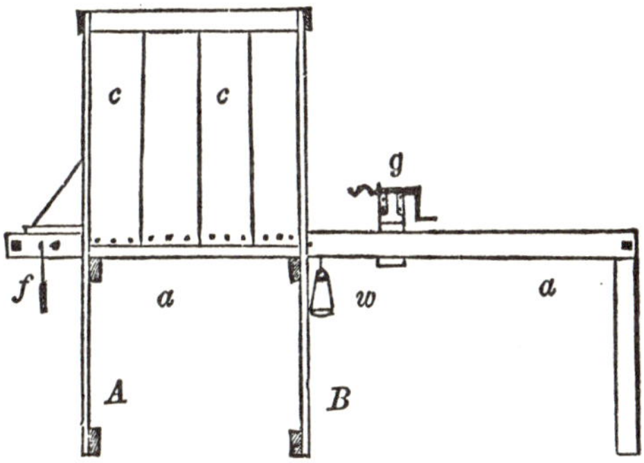

Fig. 167.—THE COMPLETE MACHINE (STRAW-TWISTER).

iron, bent over it and nailed or screwed in place; two strips are screwed on the sides, and a piece of board on the bottom end, as shown, leaving space between the two so that it will work freely on the ways, figure 163, when in place as shown in figure 167. Figure 167 shows the machine complete, *A* being figure 164, *B*, figure 165, in their places, *c*, showing boarding of rack, the edges of which show in figure 164 and 165; it is nailed to the upper piece and to the ways. *W* is a weight to bring the follower, figure 166, forward as fast as hay is used out, and keeps the hay firm and in its place against the wooden springs; the weight may be a stone, or box filled with iron or scraps. In use, draw the follower back, and run the pin, *f*, in hole bored through ways and follower, fill in the rack from top with hay, just mowed or slightly damp, pressing it in snugly, then draw out pin *f*, slide carriage *g* up to rack ; by turning crank the hook will catch up a lock of the hay, then keep turning and drawing carriage away at the same time, and it will twist out a rope of hay, this is doubled, and ends fastened by crowding through loops. If you can get some drawer rollers to set in the follower and on the carriage, it will work much easier.

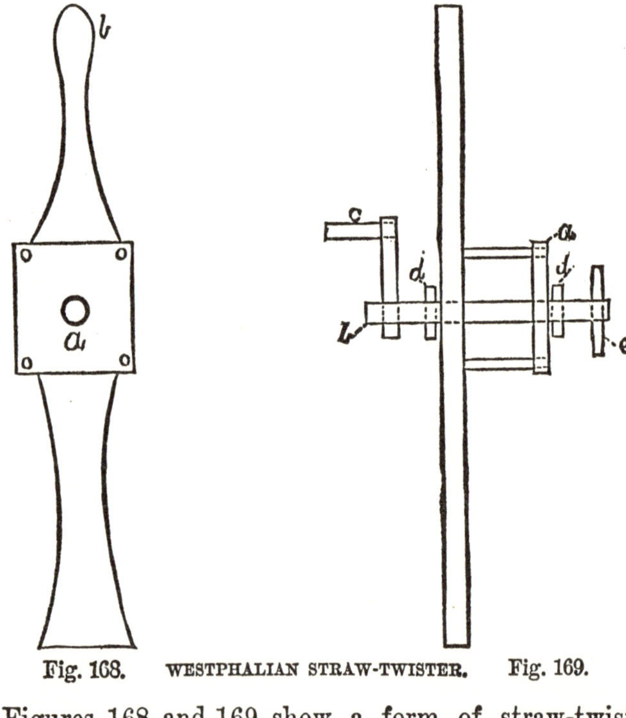

Fig. 168. WESTPHALIAN STRAW-TWISTER. Fig. 169.

Figures 168 and 169 show a form of straw-twister, which has long been in use in Westphalia, Germany. A board of hard wood, half an inch thick, four and a half inches wide and five feet long, is shaped as in figure 168, and has an inch hole in the center. A piece of the same stuff, five inches square, also with an inch hole (*a* in the engravings), is attached to the large piece by means of long wooden pins, and stands three inches from it, the holes in the two exactly corresponding. A shaft of hard wood, *b*, figure 169, is square at the end, where the crank *c*, is attached, and round where it passes through the two holes. This shaft is held in place by the pins, *d*, *d*, which should be so set as to allow it to turn freely. Another wooden pin, *e*, passes through the shaft and projects an inch and a half on each side. In making the rope, two men are required. The straw having been slightly moistened by sprinkling from a watering can, one takes a bunch and fastens it to the pin *e*. The other operator takes the machine with his left hand, at *b*, figure 168, and with the lower end against his foot, turns the crank; he gradually moves away from the other, pushing the machine along with his foot, while keeping the crank in motion, and the other supplies straw as required. The latter sits upon a low stool, and his right hand should be protected by a stout leather glove or a piece of leather. When the rope is about a hundred feet long, or it becomes difficult to turn the crank, it is rolled into a ball, and a new one begun. The rope is finally made into balls of convenient size.

STANDARD FOR CORN SHOCKS.

The best standard is made by bending four hills together—two diagonal hills being lapped and twisted together. But to such a standard it can be objected, that a knife must be carried along to cut the stalks loose when the fodder is brought in from the fields, and that these stalks cannot be stood straight in the rick. Some prefer a movable wooden standard, of which a very good sort is shown in figure 170. A light pole, twelve or fifteen feet long, is provided with two upright supports; holes are bored through the pole about five feet from one end, and through the ends of the uprights, and a bolt passed through the holes and secured by a nut. The holes should be so large that the uprights can be spread a foot apart at the bottom. Midway between the uprights and the end of the pole, another hole is bored, through which a cross-bar is put. In the four angles formed by

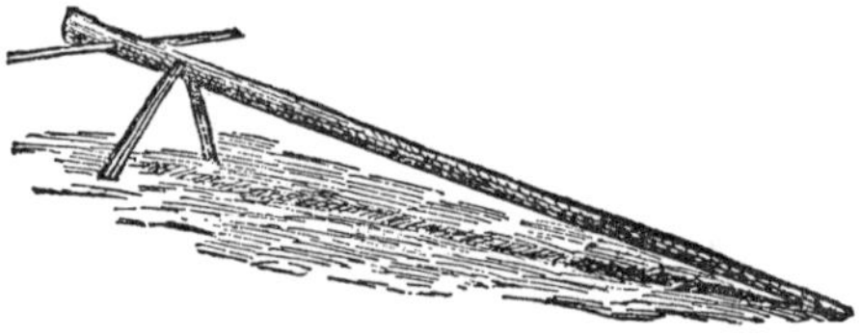

Fig. 170.—FODDER CORN STANDARD.

the intersection of the pole and the cross-bar, the fodder is set. When the shock reaches out to the support, the cross-bar is pulled out, and the pole can be removed. Some prefer to have the supports and cross-bar near together, about four feet from the pole. The shock is built around the supports. When done, the cross-bar is pulled out, and as the pole is removed, the supports are brought close together, and do not hinder.

VENTILATOR FOR STACKS.

A large quantity of corn fodder is spoiled for want of proper care in drying and stacking. It is not easy to hit the happy mean, between the sufficient drying of the stalks, and the over-drying of the leaves. But it can be done perfectly in the stack, by the use of the ventilator, figure 171. This consists of three or four poles or bars, fastened together with cross-slats, and made to fit one upon another. Such a ventilator, which may be four or six feet long, is set on the foundation for the stack, and passes upwards through it, leaving a perfect chimney and air passage in the center of the fodder. More than

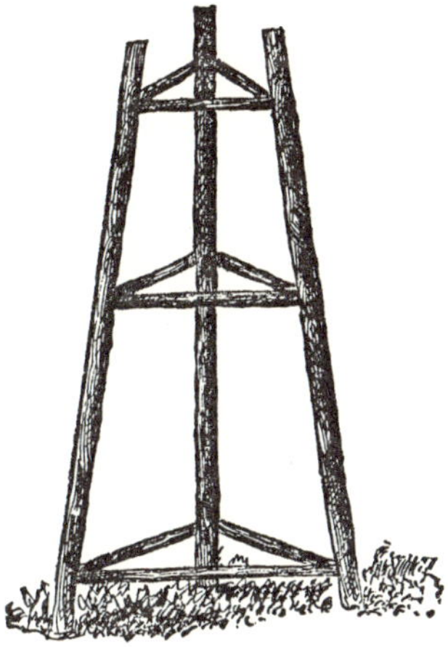

Fig. 171.—VENTILATOR FOR STACK.

one can be used if desired. These ventilators are useful in stacks of hay or grain, which may be a little damp.

BENCH FOR HUSKING.

Figure 172 represents a very comfortable and light husking stool; it is made long and wide enough to hold

Fig. 172.—HUSKING STOOL.

a sheaf of stalk, and is provided with a seat, or may have one on both sides, if desired. If one can procure some crooks of cedar or other light wood, such as is used in rustic work, they will serve very well for the ends. The seats may be removed when the stools are not required, and may then serve for benches in the dairy or for household purposes.

CORN-STALK BAND.

Stalks altogether dry or altogether green are not to be selected, as they will break when it is attempted to twist them. Long, slender stalks are desirable. The first stalk is broken at a right angle about two feet from the but; the but is then forced into the shock as far as the break, when the remainder of the stalk is passed

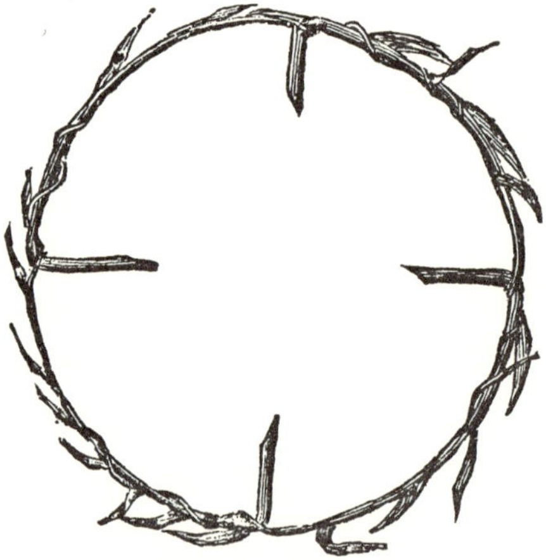

Fig. 173.—CORN-STALK BAND.

around the shock, breaking it carefully every eight or ten inches, until the tassel point is almost reached. Then another stalk is inserted in the shock. The top of the first stalk is broken every three inches between the thumb and fingers, and twisted around the second stalk, which is then broken and passed around the shock as in the case of the first one. This is continued until the last stalk reaches the first one, when it is secured by twisting it as in the case of the others, or by drawing it down between the shock and the first stalk, just in front of the break, forming a loop below, through which a piece of stalk, two feet long, is passed and driven into the shock. In figure 173 is shown the appearance of the band as it would be, if the shock could be removed after the band is completed.

CONVENIENT FODDER CARRIER.

On farms where the corn-stalks are left in the field to be carted to the yard as wanted, the use of a convenient carrier saves much work and time. Such a one is shown

Fig. 174.—FODDER CARRIER.

in figure 174. It consists of the front wheels, axle, bolster and pole of a common farm-wagon, with the ends of two poles, or a common cord-wood rack fastened to the bolster. The other ends of the poles drag on the ground. A cross-piece, three feet long, is securely fastened to the poles about three feet from their lower ends, and two upright stakes, four or five feet long, complete the arrangement of this farm convenience.

CHAPTER VII.

STUMP-PULLERS, DERRICKS AND SLINGS.

STUMP-PULLERS.

Figure 175 shows a very powerful machine for pulling stumps. The woodwork is made of well-seasoned oak, the winding shaft being eight inches in diameter and five feet long. The lower block, in which it revolves, is sixteen inches square and three inches thick, having a hole cut just large enough to receive the winding shaft, and is fastened securely to the middle brace at the bottom. To prevent the splitting of the winding shaft, two stout iron bands are shrunk immediately above and below where the lever or sweep is inserted. An old gear-wheel, with the spokes knocked out, is fastened to the top cross-piece or head-block, to receive the traveling ratchet attached to the shaft. The upright pieces of the frame are of two by eight inch oak, three and a half feet high ; the top cross-piece or head-block two by sixteen inch oak, narrowing to twelve inches at the ends, and three feet long. The frame is set on runners four feet long, two by ten inch oak, so the implement can be quickly moved from place to place ; the entire frame is mortised together. The anchor is of one-inch round iron, and attached as shown in the illustration, and a strong iron pulley-block is used on the opposite side. In pulling large stumps, a chain is more reliable than a rope. A single horse furnishes the motive power at the end of the lever or sweep, which is ten feet long.

Figure 176 shows a cheaper and lighter stump-puller. The only expense is for the chain, links of one and a half to two inch tough iron, or tough-tempered steel ; ring, ten to twelve inches in diameter, and the hook, all of

Fig. 175.—HOME-MADE STUMP-PULLER.

which any blacksmith can make. The point of the hook must be formed so that it will strike in toward the heart of the stump and not tear loose on partially decayed wood. The lever may be twelve to twenty feet long, its size depending on the quality of the wood and the force to han-

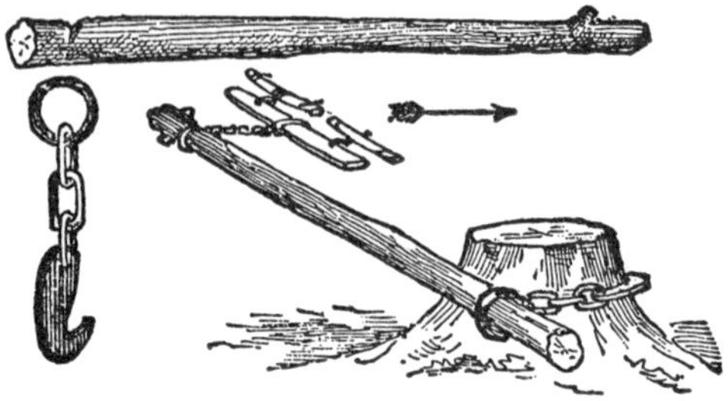

Fig. 176.—A SIMPLE STUMP-PULLER.

dle it. A lever twenty feet long on a stump two feet in diameter, would exert a force of ten tons for each one thousand pounds of direct pull by the team. Though many durable, long-rooted stumps would not yield to this, the large majority of ordinary stumps, after decaying a year or two, can thus be cleared out, with most of the roots.

Figure 177 shows a stump-puller used in New Zealand. The thread of the screw works both ways and gradually draws each chain nearer the center, where the screw is

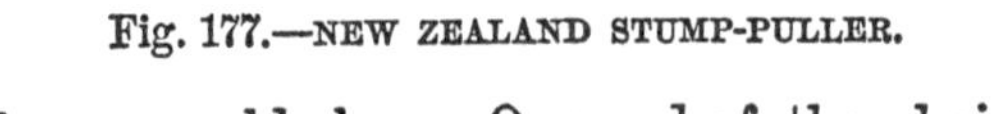

Fig. 177.—NEW ZEALAND STUMP-PULLER.

turned by a movable bar. One end of the chain is fastened around one stump, and the other around a second; then when the screw is turned, whichever stump is the less firm in the ground is bound to be pulled out. The screw is readily worked by a man, though it will, as a rule, require two persons to work it on heavy land.

DERRICKS FOR FARM USE.

Where there is much handling of heavy barrels or sacks, one man, with some simple, mechanical contrivance, can easily do the work of two or three, working by main strength. A boom derrick, figure 178, hung high, so that the weight shall be lifted from the ground ordinarily, when the derrick swings horizontally, is very convenient. A post is banded, and has a strong dowel at each end. The lower dowel is set in a stone fixed in the ground, close to the building where it is to be used, the

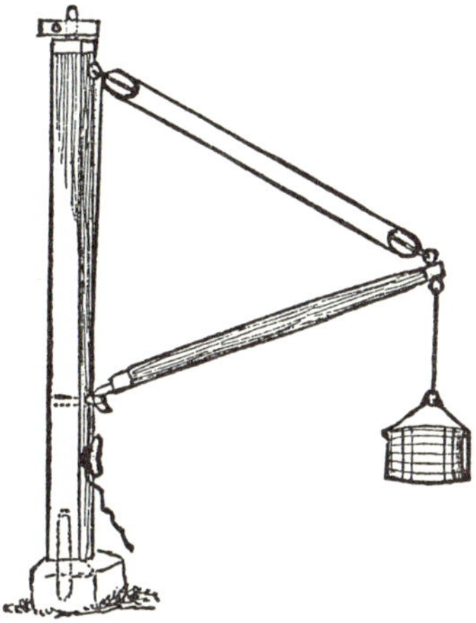

Fig. 178.—A BOOM DERRICK.

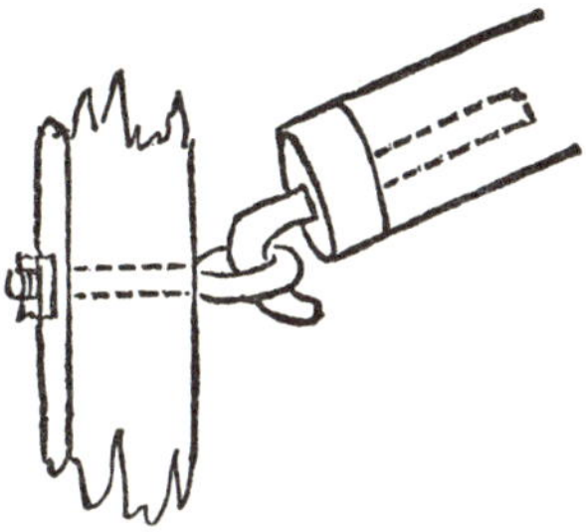

Fig. 179.—EYE-BOLT.

upper one in a strong oak cleat, bolted to the building. At the height of about five or six feet from the ground, an eye-bolt passes through the post, and another is fixed at the top. The boom is fastened to the lower eye-bolt by a three-quarter inch hooked iron, attached as shown in figure 179, while the other end of the boom has a band with two eyes. This boom is a spar or pole, stiff enough to bear the strain without doubling up or breaking, and may be ten or fifteen feet long. The end of the boom is raised or lowered by a pair of single pulleys, or by a double block tackle, which will exert much greater power. When the weight is lifted, as out of a cellar-way, it may be swung around over a wagon and lowered into it.

A convenient derrick for raising slaughtered animals, for suspending heavy hogs in scalding, and dressing beeves, and for sundry other purposes, can be cheaply and quickly made thus: Take three scantlings two by six inches, and fourteen feet long, or any other desired length and strength. Round poles will answer, by hewing flat on two sides a small portion of the upper ends. Bore corresponding holes in the top of each, and insert a strong iron bolt, with large head on one end, and large nut and screw on the other. Let the bolt fit loosely, to allow a little play. These pieces can fold together for storage, and be raised to any desired height short of perpendicular. Bore a series of small holes along the upper sides of two poles, for movable iron pins, or larger ones for wooden pins. These may be fastened in, or better, have two loose pins for moving to higher or lower holes. By placing the feet of these two poles against firmly driven stakes, and drawing the third and rear pole inward, the center will be elevated with considerable force, the power required decreasing as the timbers approach a perpendicular, when a beef carcass, for instance, is nearly lifted from the ground, and hangs more heavily. If desired or necessary, horse power can be applied by using a rope with a clevis or otherwise, attaching it to a double-tree or to a whiffletree. A single horse will be sufficient for raising a large carcass by means of this tripod derrick.

SLINGS FOR HOISTING HEAVY OBJECTS.

When one has bags to hoist by a block, or simply by a fall, from the barn floor to the loft, rope or chain slings are almost essential. The simplest sling to operate is formed on the end of the fall-rope, as shown in figure 180. This consists simply of an oak stick, half an inch

Fig. 180.

Fig. 181.—CHAIN-SLING.

thick, two inches wide and six inches long, having two three-quarter inch holes bored, one near each end. Through one of these the end of the rope passes, then it

Fig. 182.—ENDLESS ROPE-SLING.

is drawn through the other and knotted strongly. The mouth of the bag being caught in the bight of the loop, it may be safely hoisted, for the greater the weight the tighter will be the hold.

Next to this, and still more convenient, is the chain-sling, figure 181. The fall-rope is terminated by a chain with twisted links, which ends in a ring, and so a loop is made to take the bag, or simply the bag's mouth. Like the rope-sling, it will hold fast all the bags that it can be made to surround. For hoisting many bags at a time, nothing is more convenient and safe than an endless rope, figure 182, cut eighteen to twenty feet long, and the ends spliced together. This is laid upon the floor, forming a long, narrow loop; the bags are laid upon it, resting evenly on both side ropes, then the ends are brought together, one is passed through the other, so as to act like a noose, and hooked over the fall-rope, which should terminate in a strong hook, as shown in the engraving.

DERRICK FOR A CELLAR.

The carrier shown in figure 183 is similar to those used for hay, but more simple in construction. Four iron wheels are attached with bolts, which serve as axles, to two-by-four-inch oak blocks. The connecting bars holding the blocks together are made of old wagon-wheel tire, and joined together below the carrier by a cross-bar of the same material, bearing a hook. The track is made of a bent two-by-four-inch scantling, to each side of which are bolted oak strips one inch thick, forming a roadway for the wheels. To the outer end of the track is fastened a pulley, over which passes the rope attached to the carrier. When the lead runs into the cellar the rope moves along in the groove under the track. In removing heavy articles from the cellar, the end of the rope is attached to a windlass, set a short distance from the cellar door. When not in use, the carrier may be taken down and laid aside out of the way. This device has

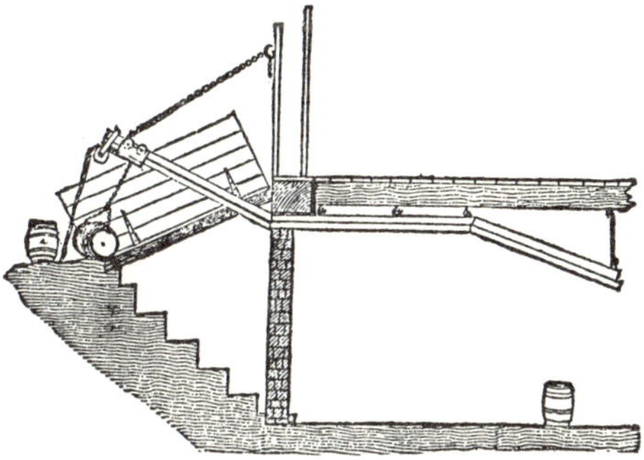

Fig. 183.—A CELLAR CARRIER.

been used in a cellar, where four hundred bushels of apples and potatoes were stored, and the owner would not part with it, if he could not obtain another. With it one man can place a hundred barrels of apples in a cellar, or remove them, as easily as he could as many pecks with his hands alone.

LEVER APPARATUS FOR LIFTING.

The implement shown in figure 184 is very useful for many purposes on the farm. Mortise a post of three by

Fig. 184.—CONVENIENT LIFTING APPARATUS.

three-inch stuff into a piece of two-inch plank. In the top of this saw a slot, one and a half inch wide, to receive the lever of the same thickness, four inches wide, and with the short arm, three feet long, and the long arm, six feet long. To the long arm is fastened a piece of chain, and to the short arm another piece, provided with a hook at the free end. Having the long arm of the lever twice as long as the short arm, one can easily lift a weight twice his own. It is surprising how often there is use for this. It can be used to lift sacks of grain into the wagon; logs on the sled or saw-horse; the bed off the wagon; the mower over an obstruction when putting

Fig. 185.—IMPROVED LIFTING APPARATUS.

it in the barn; and for some other things nearly every other day. By making the chain on the short lever long enough, it can be passed around a log or sack, and hooked very quickly.

The improvement shown in figure 185 consists in having the long arm of the lever longer and the short arm a very little shorter, giving a greater advantage. As the short arm of the lever is brought up, the free end of the chain is shortened; hence, it will lift the weight a greater height. With the first device one can lift a weight only three feet conveniently.

A HOME-MADE HORSE-POWER.

The worst disposition which can be made of a worn-out farm machine of any kind, is to lay it up by the fence in the dooryard or barnyard, to be an eyesore for years, and a possible source of danger to domestic animals, or even to persons passing hastily or carelessly near it. The better plan is to take the machine to pieces, set aside any unsound or broken wood for fuel, sell whatever iron is not likely to be useful in its present shape, and carefully store away in a suitable place the remaining parts, whether of wood or iron, particularly bolts, gearing, etc. With a little ingenuity, and perhaps a slight outlay of money, wheels and shafts from disabled reapers, mowers or other machines may be put together to form a light horse-power, which will be found very serviceable in driving feed-cutter, corn-sheller, or farm-mill. In the construction of a horse-power certain general principles must be kept in mind, otherwise failure, more or less complete, will be the result. The different parts must be sufficiently strong to bear the strain to which they will be subjected ; the bearings need to be true, and the whole so securely braced and held together, that any slipping of cogs will be impossible. The rate of speed must be from seventy-two to one hundred and sixty revolutions of the cutting-box shaft for every one of the horse, the first being rather low for a six-foot, and the second rather high for an eight-foot sweep. Since some portion of the force employed is always lost through friction, the fewer wheels to secure the required speed and direction, the better. Hard-wood boxes are cheaper and are more easily adjusted than those made of metal, and, if they are kept properly greased, last, perhaps, quite as long.

An excellent portable-power can be made by taking a bevel-gearing from an old discarded brick-machine, a pair of spur-wheels from an ancient reaper, two or three shafts and a band-wheel from other sources—all odds and ends picked up cheaply here and there—arranging them to suit the purpose, and fitting all but the band-wheel and one shaft in a stout frame. The odd shaft extends from the end of the frame some distance, and carries the band-wheel at its further end, above which a feed-cutter stands on a loft, and is run by a belt. From a pulley on the same shaft, power is conveyed to a grindstone and corn-sheller, which require a much lower rate of speed than the cutter. The crown wheel has fifty-four cogs, its pinion, eighteen; the spur-wheel has seventy-two cogs, its pinion, fifteen; the band-wheel is thirty-six inches in diameter, and the pulleys on the cutting box, six inches. The number of revolutions of the cutter-shaft to one of the horse are, therefore, eighty-six and two-fifths. A six-inch leather belt will seldom or never slip; a four-inch belt is quite too light. Two horses, attached to this power, cut cornstalks very rapidly. The crown-wheel has a tendency to rise and allow the cogs to slip. It must be kept down by friction wheels placed above the rim, or by a collar on the axle, working against the underside of the upper cross-piece, which, in turn, must be kept in place by a bolt or rod at each end, running up through the bed-piece, and secured at the top by means of a broad washer and stout nut.

CHAPTER VIII.

PREPARING AND HANDLING FERTILIZERS.

HAULING BARNYARD MANURE.

When hauling manure it is usual to drop it in heaps, and leave it to be spread by a man who follows soon after. There are several methods of dumping the manure, but the most satisfactory is to use a manure hook, as shown in figure 186. The bottom of the sled or wagon should be formed of loose planks, each with its end shaved

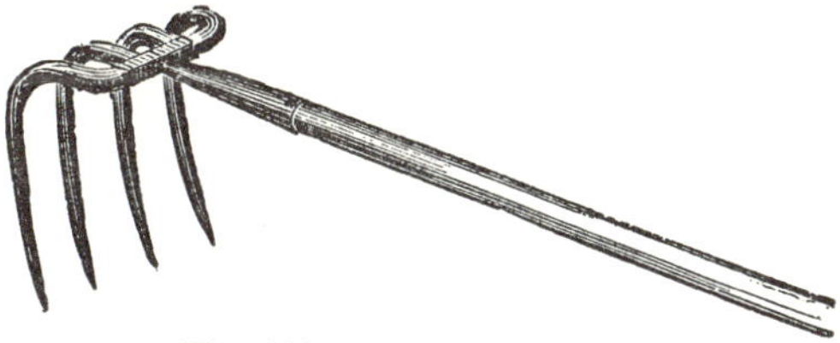

Fig. 186.—A MANURE HOOK.

down to form handles. The side and end pieces of the box, though closely fitting, are not fastened together, so that they can be removed one at a time. One side or an end board is first taken out, and with a manure hook a sufficient amount of the load removed for the first heap. The manner of unloading the manure from the box

Fig. 187.—A MANURE WAGON BOX.

above described, is shown in figure 187. The other side and ends are afterwards taken off, and finally the bottom pieces are raised and the sled or wagon is soon emptied. In dropping the heaps, they should be left, as nearly as may be, in straight rows, and of a size and distance apart determined by the amount of manure to be spread. If they are placed regularly one rod from another each way, and eight heaps are made from a load, there will be twenty loads per acre. In spreading such heaps the manure is thrown eight feet each way, and the whole ground is covered. It is important that the spreading be done in a careful and thorough manner, each portion of the surface getting its proper share of the manure. It is important also that all lumps be broken up.

IMPLEMENT FOR FINING MANURE.

It is often desirable to have fine manure for use in hills and drills; and it is also at times necessary, when artificial fertilizers are lumpy, to pulverize them for use. A tool for this purpose is shown in figure 188. This is

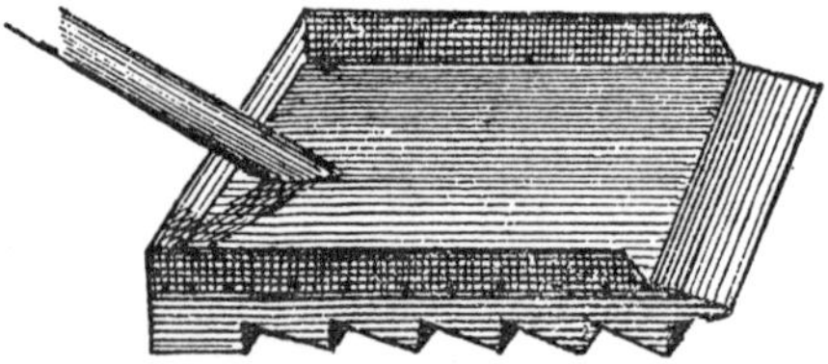

Fig. 188.—IMPLEMENTS FOR FINING MANURE.

especially useful in preparing the mixture of poultry manure and plaster. The implement is made of a piece of three-inch hard-wood plank, twelve inches wide, sawed and cut across into notches, and surrounded on three sides as shown, with a strip of sheet iron, or broad hoop-iron band. It is rubbed back and forth over the manure on a floor, and can be used as a shovel, by raising the handle, for turning over and mixing the mass.

MUCK AND PEAT.

Fresh muck contains valuable plant-food, but usually in an unavailable form. There are many instances where muck, applied to land, has proved positively injurious. Muck needs to be exposed to the action of the frost, rain and sun, or, as it is termed, "weathered," for a season, before it is fit to be used as a fertilizer. Even after it has thus been subjected to the elements, it is usually best to employ the finely divided muck as an absorbent of liquid manure in the stable or shed, or even the barn-yard. In this way the food elements are brought into a better state for the plants to feed upon. If the "weathered" muck and manure can be composted together for a time, a still more valuable fertilizer is obtained.

When one has peat or muck in any form upon his farm, it should, of course, be dug when the water is low in the swamps, and the task of getting out muck may

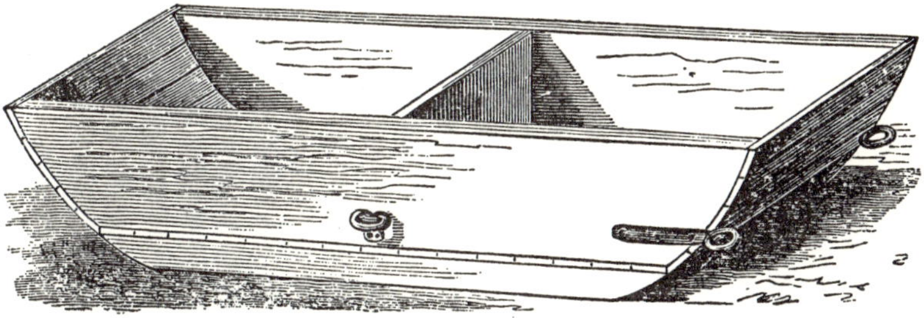

Fig. 189.—A BOAT FOR GETTING OUT MUCK.

aid essentially the work of reclaiming the swamps. Thus the main ditch may be dug the width of a cart track. By making a narrow preliminary ditch to carry off the water and dry the ground, a horse and cart may be brought into the ditch and the muck carted directly off to dry ground, where it can dry, and perhaps be exposed to a winter's freezing and thawing, before using in the compost heaps or barn-yard. In all such ditching we must begin at the lowest end of the ditch, so that there shall always be a free outlet for the water. A boat, to be used in removing muck from the bed through a water channel to a hill-side, is shown in figure 189. It is of pine boards, nailed firmly to side planks, braced by a cross plank at the middle. If made nine feet long, four feet wide, and sixteen inches deep, it will float a ton of muck. A runner is placed under each side, so that the boat can be drawn upon the land. A hook or eye should be placed on each side, and others at one end, by which the boat may be drawn. While floating, the boat is moved by handspikes. The place where the muck is heaped to dry, should be as near as possible to the bed from which it is dug.

The muck may be very peaty, or the material really may be *peat*—that is, consisting almost entirely of vegetable matter and ash—whereas *muck*, as the word is applied in the United States, is used to mean such as would be of little or no value as fuel, from the amount of soil or sand or calcareous matter in it ; but it is useful as manure. The peaty mucks are greatly benefited by being treated with lime—in fact it is only by acting upon them with lime or ashes that they can be made rapidly fit for composts or for application to the land. The old rule to slake stone-lime with strong brine, adding only brine enough to dry-slake the lime, is a very good one. Such lime may be depended upon for the best results when composted with muck.

HOW TO BURN LIME.

The application of lime improves the mechanical texture of heavy soils, and this will frequently compensate for its use, if the lime can be obtained cheaply. In many localities, the farmer can burn the lime he needs, and thus obtain it at a much less cost than the market price. It is not necessary to build a kiln of masonry. The cheapest kiln is made by digging an excavation in a bank, as shown in figure 190. If much lime is to be burned, it will pay to line this excavation with brick, and place an iron grating across near the bottom, beneath which the fire is made. Whether the kiln is so made, or constructed only in a temporary manner, it must be banked up in front with earth, after the limestone is placed in it. Where the iron grating is used, it should

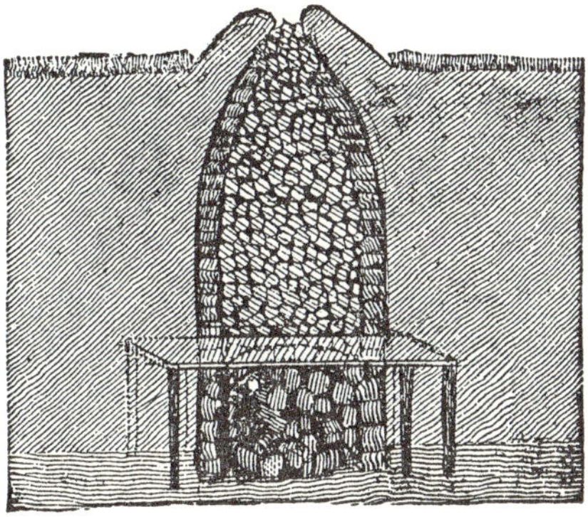

Fig. 190.—A LIME KILN.

project out in front as far as the bank of earth will permit, while under it is used a sheet iron door, to close the furnace and regulate the draft. A platform is built just above the projection of the grating, to support the earth banked against the rock. The top of the heap is covered with earth, leaving a hole in the center for a chimney. When the kiln is only temporary, an arch of large rocks takes the place of the iron grating, and the sides of the kiln are lined, as the rocks are laid in, with large stones instead of brick. It will take four or five days, with a good fire, to burn the kiln sufficiently.

Lime may also be burned by piling the stones in a

conical heap above ground. Large stones are used to make an arch under the heap, and the cavity below the arch is filled with fuel. Immediately above the arch is placed a layer of dry wood, then a layer of lime-stones, next a layer of wood, and so on until the heap is completed. The stones are laid rather loosely, and the entire heap is covered with earth to the depth of at least a foot, to retain the heat, leaving an opening at the top for the escape of smoke. It will pay to insert a short sheet-iron chimney in this opening, to increase the draft, as a hot fire is needed. The draft is regulated by opening or closing the doorway under the arch. Do not disturb the heap until it is perfectly cool, and if the lime is not to be used at once, it should be protected from rain by a roof, and from surface water by erecting a low bank about it. Where limestone boulders can be gathered in sufficient quantities, the cost of lime will be very little, and even when the rock must be quarried, burning lime will frequently yield handsome returns when the weather does not admit of regular farm work.

VALUE OF GAS LIME.

Gas lime, as its name indicates, is a product of gas works. Quick lime is spread in large boxes, called purifiers; the gas passes through these, and coming in contact with the lime is deprived of its impurities, especially the sulphur it contains. When the lime ceases to act, it is thrown out and replaced by a fresh supply. Gas lime smells strongly of sulphur, and contains the sulphides of ammonia and of lime. These are fatal to plant life, and before it can be used as a fertilizer, the lime must be exposed to the air for some weeks. When unpleasant odors are no longer perceptible, the gas lime may be used in the same manner as ordinary lime. It still consists largely of quick lime, and contains more or less sulphate of lime (gypsum or plaster), formed by the conversion of the dangerous sulphide into sulphate of lime. It may be used after exposure to mix with muck, but cannot safely be used in its fresh state. It is an easy matter to expose it before adding it to the muck, and thus be on the safe side.

BURNING CLAY AND SODS.

Burning clay for manurial purposes, is an old fashion, which deserves renewed notice and practice. Along with the clay or with ordinary soil or swamp muck, may be mingled coarse sods, the scrapings of road-side ditches,

Fig. 191.—BURNING CLAY AND SODS.

the mossy surface and hard tussocks of swamp meadows, rough "waste wood," coarse weeds, and other similar matters which slowly decay, and are of no value until they are decomposed. These combustible matters are placed in small heaps over an old meadow, which needs renewal, or any other piece of land. The rough waste matters being gathered, placed, and covered with earth, so that they will burn slowly, in the manner shown in figure 191; care being taken to so arrange them, as to distribute the heat all through the mass and the earth with which it is covered. These heaps are fired and left to burn slowly for several days, when the dust and ashes are spread over the surface. The lime and potash thus made available, both from the waste material and the earth covering, furnish considerable fertilizing matter.

CONVERTING STRAW INTO MANURE.

In the West the object is to feed one-third of the straw stack, and convert the balance into manure as rapidly as possible. The straw trampled under foot by the cattle will not thoroughly rot within a year, if left to itself. To rot and fine, it must be stirred about, and the swine can be made to do this work. If the hogs are fed on the straw twice a week, they will move the entire mass, unless quite deep, rooting after stray grains. If their noses do not get to the bottom of the heap, sharpen a heavy stake and prod it through the straw ; then withdraw it and drop shelled corn or oats into the hole. In this way a hole can be made every few feet over the pile, and the hogs will turn the manure thoroughly. A hog's snout is a very cheap and effective manure hook. The hogs must not be allowed to lie on the rotting straw, as this is almost sure to produce disease among them. They become too warm, and then when they come into the open air contract colds, catarrhal or pulmonary diseases. If the hogs are used as above recommended, straw can be converted into well rotted and fined manure within six months ; and if the straw stack is put on level ground, not much will be lost during this rapid conversion. When from twelve to eighteen months are required for the rotting of the manure—and this time will be required when deep masses are not disturbed—and the straw is on a side hill, not a little of the value of the manure is lost by being washed down hill.

MANURE FROM MARL AND SHELLS.

Marl is quite abundant in some localities, and in others oyster shells can be had for the hauling. These may be profitably burned into lime for use as a fertilizer. In regard to this use of lime, it should be remembered that the

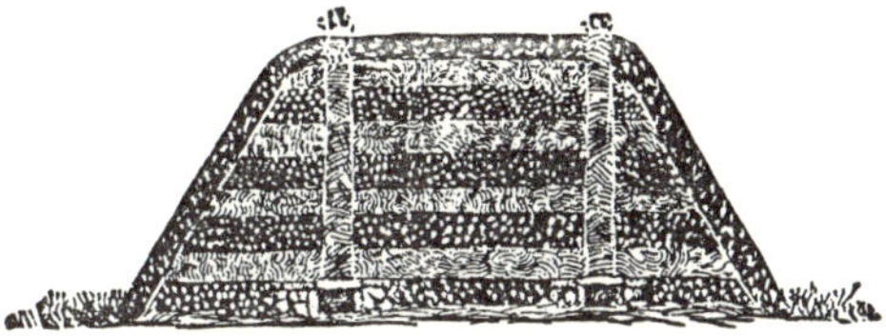

Fig. 192.—PIT FOR BURNING MARL.

larger part of the ash of agricultural plants consists of lime, and that it is thus an indispensable plant food. It is rarely used without benefit, but is most useful when applied in a caustic state, or when it is freshly burned. Enquiries are often made about burning marl and shells. An easy way of doing this is in piles, commonly called "pits," made as shown in figure 192. A level spot is chosen, and a quantity of small wood is spread over it, either in a square, or better, in a circle. Two or three double rows of stones, covered with other flat stones, are laid as at *a*, *a*, to form flues. A layer of shells or marl is thrown upon the fuel, and other alternate layers are added, until a conical heap is made. Chimneys of small wood or chips are made over the flues as the heap is built, and carried to the top.

MAKING FERTILIZER FROM BONES.

It is well enough known that bone, when ground fine, makes one of the best and cheapest manures, especially on lands long in use. The needs of farmers with abundant capital are well enough met in the commercial fertilizers. With the Experiment Stations to analyze the samples, there is not much danger of adulteration. The high price of this comminuted bone, two cents a pound and upward, deters many farmers from using it on a large scale, even where there is no doubt that the investment would pay. In a limited way, the small farmer has the means within his reach, of reducing several barrels of bones to a fine powder every year. A solution of potash will reduce bone to a fine condition, and make it available for plant food. Most farmers still use wood for fuel, and the ashes from the fifteen or twenty cords used in a year, if saved, would reduce all the bones ordinarily within reach of the farmer. The old-fashioned leach that used to stand at almost every farmer's back-door for soap-making, was a good contrivance for reducing the bones. But any tight, strong cask or box, will answer quite as well for this purpose. Water poured upon the ashes makes a lye, or solution of potash, strong enough to decompose the bones. The casks should stand under cover, so that the quantity of water applied to the bone and ashes will be under control. The time it will take to reduce the bone to a powder, will depend upon the amount of potash in the ashes, and attention bestowed upon the process. It is essential that the ashes and bone should be closely packed in the mass, and that they be kept in a moist state, adding water as it evaporates from the surface. The finer the bone before it is packed in the ashes, the sooner will it be reduced. The process can be hastened by putting into the mass a few pounds of common potash. But this is only necessary to save time. Ashes from hickory or any other hard wood contain sufficient potash to decompose the bone. When the mass is soft enough to break down with a spade or shovel, it can be mixed with land plaster, dried peat, or loam, to make it convenient for handling. It is a concentrated fertilizer, to be used with discretion in the hill, or applied as a top dressing to growing crops in the garden or field. We are quite sure that any one who uses this preparation of bone and wood ashes, and sees the vigorous push it gives to garden and other crops, will be likely to continue it. But many farmers near seaports and railroad stations, use coal mainly for fuel, and will have to resort to a hand or horse-mill to use up the waste bones. Small mills are extensively used by poultry-men, for crushing oyster shells as well as bone, and the machine can be adjusted to break the bone coarsely for hen feed. The oil and gelatine of the bones have an alimentary value, and, turned into eggs, pay much better than when used as a fertilizer for the soil.

CHAPTER IX.

APPLIANCES FOR THE GARDEN AND ORCHARD.

PAPER PLANT PROTECTOR.

The most effectual means for protecting young melon and cucumber plants against some of their injurious

Fig. 193.—PATTERN FOR PLANT PROTECTOR.

enemies, is to inclose the young vines in bottomless boxes of some kind. Various more or less expensive and elaborate forms have been invented and are offered for sale. The principal objection to most of these is their cost.

Figures 193 and 194 represent a device which is free from this objection. It consists simply of a piece of card-board

Fig. 194.—PLANT PROTECTOR.

or stiff paper of any kind, as seen in figure 193. When the ends are brought together, and the slits, indicated in the engraving, made to interlock, a cone, as seen in figure 194, is produced which, when put around a plant, furnishes as complete a protector against insects as the most expensive device.

MUSLIN-COVERED PLANT SCREEN.

To make the device, figure 195, take four strips, one-half inch thick and one inch wide, and twelve

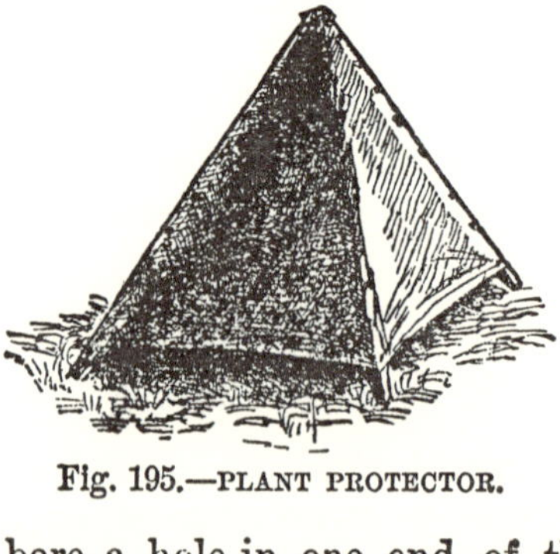

Fig. 195.—PLANT PROTECTOR.

inches long; bore a hole in one end of these, through which pass a wire, the ends of which are twisted together, but not so tightly as to prevent the opposite ends of the pieces from being spread apart from eight inches to a foot, making a tent-shaped frame. Cheap muslin is tacked on the frame, spreading the pieces before doing so. The muslin should be brought down to within about two inches of the ends of the sticks, so as to allow them to be run into the ground that distance, when in use. When not in use, the protectors can be closed up and take but little room, and if properly cared for, they will last several seasons.

PROTECTED PLANT LABEL.

Various devices to prevent the washing off of the names written on plant labels have been invented from time to

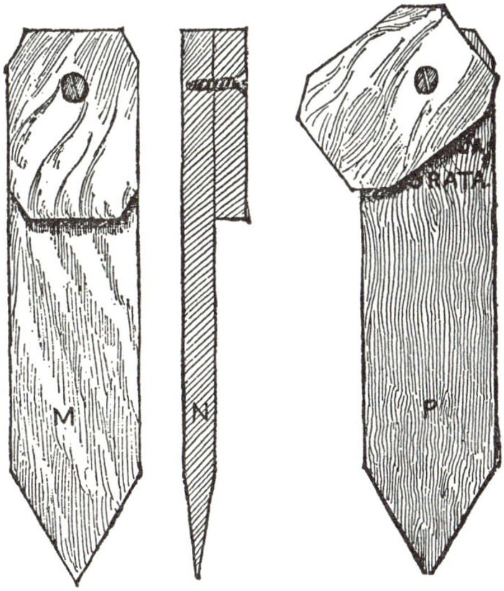

Fig. 196.—IMPROVED PLANT LABEL.

time. A novel one is shown in figure 196. It consists in fastening, with a small screw, a short piece of label over the name, as seen at *M;* a cross-section of the label is shown at *N*, and at *P* the cover is partly raised. This arrangement may be applied to any size of labels. The great difficulty with wooden labels is not that the name becomes obliterated, but that the portion in the ground decays. This may be avoided by using Locust, which is expensive.

POLES FOR BEANS AND OTHER CLIMBERS.

White birches and alders, so commonly used for bean poles, are about the poorest, for they last only one season at the best, and sometimes break off at the surface of the ground and let down the beautiful pyramid of green before the pods are ripe. White Cedar from the swamps is durable, and the rough bark enables the vines to climb without any help from strings, but these are not always accessible. Red Cedar is much more widely distributed, and on the whole makes the best bean pole. The wood is as durable as the White Cedar, and young trees, from which poles are made, grow quite stout at the ground, and, if well set, will resist very strong winds. A set of these poles will last for a generation. For bean poles, all the side branches are trimmed off, but for a support for ornamental climbers, these may be left on. A Cedar, six or eight feet high, with the branches gradually shortened from below upwards, makes an excellent support for ornamental vines. One of these, covered with a clematis, or other showy climber, makes a pyramid of great beauty. It is well to prepare a supply of poles for beans and other plants before the work is pressing.

POTTING STRAWBERRY PLANTS.

Figure 197 shows a simple method of turning old tin cans into contrivances for potting strawberries. Unsolder the cans, and cut into pieces of about three by seven inches. Turn back one quarter of an inch of each end, so that when the strips are bent around, they will clasp together, as shown in the engraving. In potting, the contrivance is placed on the bed near the runner, and pressed into the ground with the sole of the foot. The sharp edges allow it to push through the soil easily. A runner is placed in the center of the cylinder, and held by a stick or stone, which also serves to mark the place. When the plants are well rooted, the tin pots are taken up, unclasped, and the ball of earth placed in the new bed provided for it.

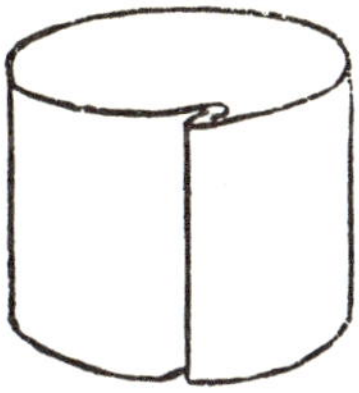

Fig. 197.
PLANT POTTER.

STAND FOR BERRY BASKETS.

During the berry picking season much time is lost in the field, through the lack of a suitable box or stand for

Fig. 198.—BERRY PICKING STAND.

transferring the filled baskets to the packing house. Figure 198 illustrates a very convenient and simple stand for this purpose. It may be made to contain either nine, twelve, sixteen, or twenty baskets, as may be desired. The handle is made of a barrel hoop nailed firmly to the sides. Suitable legs are attached to the stand to raise it from the ground.

TUBE FOR WATERING PLANTS.

Figure 199 shows an implement for watering garden plants. It is a tin tube, one-half inch in diameter, eight inches long, perforated near the bottom, and with a conical end. The upper end, *b*, is in the form of a funnel. In using this device, insert the conical end of the tube in the ground as near the plant as convenient, without disturbing the roots, and turn the water into the funnel. The water will pass out into the soil through the perforations at the bottom. The soil is not baked on the surface when watered in this manner, and the operation is very quickly done. Any local tinsmith can make the tube at a slight expense. A small flower pot is sometimes sunk in the soil near the plant, and the water, when poured into it, will gradually soak away.

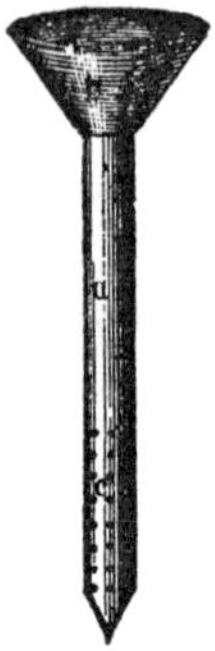

Fig. 199.

MOVABLE TRELLIS FOR GRAPES.

A grape trellis, possessing several good points, is shown in figure 200. The wooden posts, which need not be fastened together, are of 3 by 4 stuff. If leant against each other, their own weight and that of the vines will hold them in place. They are joined by smooth galvanized fencing wire. The posts must be braced inside, as seen in the illustration. If it is desired to lay down the vines in the fall, the staples can be drawn and the wires pulled out, greatly simplifying the work. The trellis being double, a row of vines may be planted on each side. Another point in its favor is that it allows the picker to get at the bunches on the under sides of the vines easily and without disturbing the vines. The trellis is as cheap as any, is strong and durable, and does not require the

Fig. 200.—MOVABLE GRAPE TRELLIS.

digging of holes, while it may be taken out of the way when the vines are laid down, and stored under shelter in a small space until needed again.

TOOL FOR CUTTING EDGINGS.

Fig. 201.

No grass-plot, walk, or flower-bed, set in turf, is complete, unless its edges are kept neatly trimmed. The work may be done with a sharpened spade, but it is better to use a regular edging-tool. An old hoe can be taken to the blacksmith, who will straighten out the shank, and round off the corners of the blade with a file, and the tool is an excellent sod-trimmer, and very light to handle. For cutting sods, this makes much easier work than the spade, though that is needed for lifting the turf when cut. Using a board as a guide, the cutting will be rapidly done.

SUBSTITUTE FOR PEA BRUSH.

The best substitute for pea brush is a trellis of galvanized iron wire. The peas are sown in double rows, six inches apart. A post, six inches in diameter, is set firmly at each end of the row; it may be round, set three feet in the ground, and of a hight suited to the variety of pea. As soon as the vines are large enough, the wire is made fast to the post, about six inches from the ground, carried to and passed around the post at the other end, and back to the starting point. Here it is made fast; it may be cut off, but still better, two or three turns are taken around the post and another double wire stretched about eight inches above the first, and so on until as many wires as needed are put in place. No. 18 wire, which measures 150 feet to the pound, is suitable. If over 200 feet long, a similar post should be set mid-way of the row. Stakes (plasterers' laths will answer) are set every ten or fifteen feet along the row, to keep the wires from sagging. These have notches cut in them, in which the wires rest; or the wires may be attached to them by means of staples or cord. When no longer needed, the wire is wound up on a reel, and, with the posts, stored away for another year. Pea-growers for market allow the vines to lie upon the ground, and claim that the crop is not enough larger when brushed, to pay the cost of cutting and placing the sticks. In the garden, neatness, and especially the greater ease of picking, make it necessary to use brush, or a substitute. The chief precaution to be observed is, to have the wires of this trellis so near together that the vines can reach them as soon as a support is needed.

TRELLIS FOR TOMATOES.

A tomato trellis, which never fails to give satisfaction, is shown in figures 202 and 203. The standards or legs are made of one by one and a half inch stuff, three feet long, and tapering slightly toward the top. The slats are selected lath. Figure 202 is an end view of the trellis in position ; figure 203 shows the trellis folded. Wires

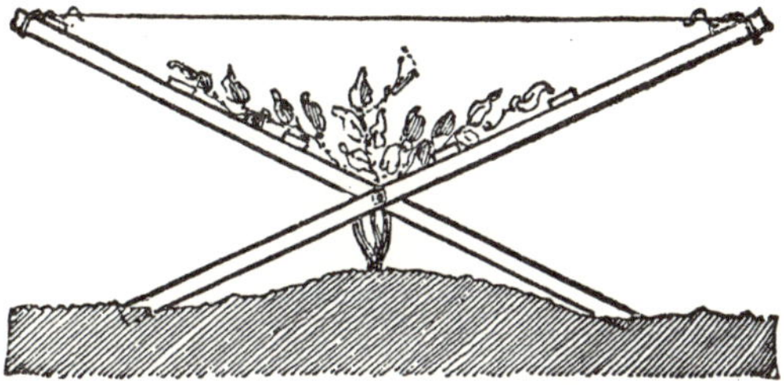

Fig. 202.—END VIEW OF TRELLIS.

extend across the top of the trellis, and when in position, they loop over the ends of the stands, and hold it at the proper width. The standards are fastened together where they cross with one-quarter inch bolts, two inches long. Two lengths of the trellis are sufficient for three tomato plants. It may be placed in position when the plants have attained a hight of six or eight inches. At the end

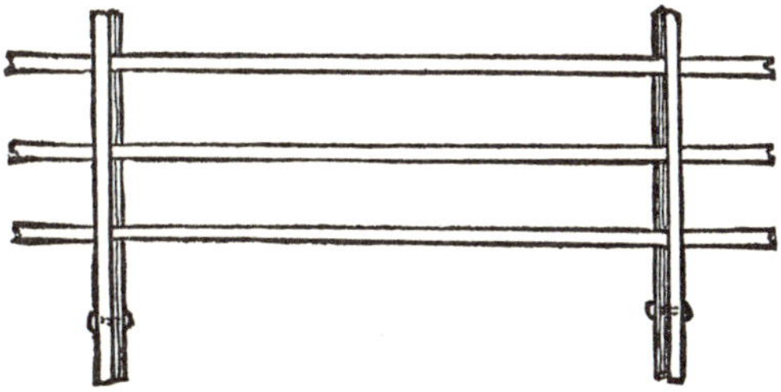

Fig. 203.—THE TRELLIS FOLDED.

of each season, after the crop is gathered, the trellises are taken up, given a coat of paint or crude petroleum, folded and packed away in a dry place. This form of trellis has the advantages of strength, lightness and portability.

TOOLS FOR KILLING WEEDS.

Weeds are easily killed when they are first seen, and more easily still, before they are seen at all. A heavy rake is better than a hoe for this work, and will do more in ten minutes, than can be done with a hoe in an hour. An implement made as in figure 204, will do this work of

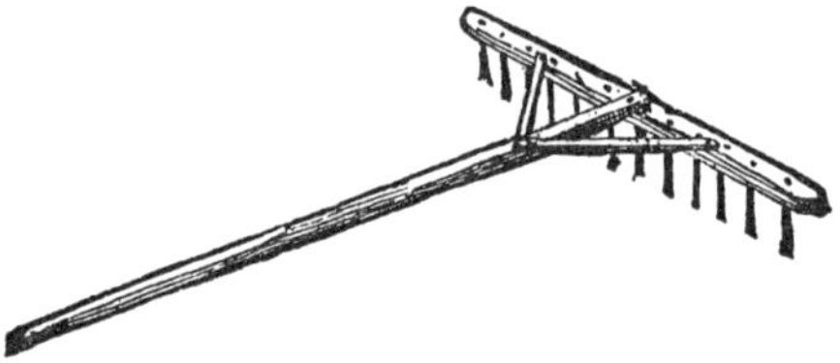

Fig. 204.—RAKE FOR WEEDING.

weeding in an excellent manner. This is made of a heavy rake head, with a handle attached as shown, and furnished with a number of teeth placed about an inch apart. The teeth may be made of forty-penny nails, or one-quarter inch round iron, the weight of which will bury them in the soil without any effort. It is much more easy to work with this implement, than with a lighter rake. The beds may be cleaned close to the plants, and it should be used as soon as the weeds begin to appear.

For killing perennial weeds, a spud is a convenient implement with which to cut off the roots below the surface.

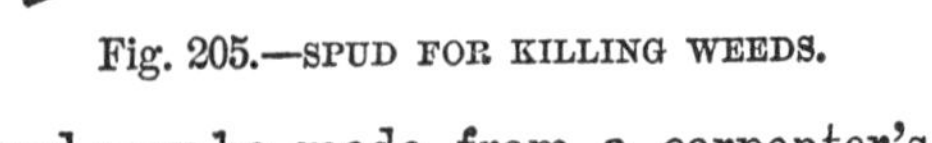

Fig. 205.—SPUD FOR KILLING WEEDS.

A good spud may be made from a carpenter's chisel of large size. This should be attached to a handle sufficiently long to allow it to be used without stooping. By thrusting this diagonally against the root, that may be cut off as far below the surface as desired. Some weeds, however, such as dandelion, plantain, etc., are not killed by merely cutting them, but need the application of some destructive liquid to make complete work. In England, oil of vitroil (sulphuric acid) is used for this purpose, but that is dangerous to handle, and must be kept in glass. Strong brine or coal-oil is sometimes applied to the roots to destroy them. We give an illustration of a vessel for the application of liquids, which is attached to the spud, and allows the cutting and killing to be done at one operation. Figure 205 shows the spud, *a,* with its attachment, a tin vessel with a tapering nozzle and holding about a quart, at *b.* At *c,* is a valve, which covers a small air-hole, against which it is pressed by a spring, and which may be raised by the cord, *e.* After cutting the root, a pull of the cord will raise the valve, allow air to enter the vessel, and a small quantity of the liquid will pass out and come in contact with the root.

VARIOUS FRUIT PICKERS.

A good picker is shown in figures 206, 207 and 208. Figure 206 is the picker. The pieces, *a* and *b,* are iron, shaped as seen in the cut. They work on a rivet, and are fastened securely to the end of the pole. Holes are punched through *a* and *b,* and stiff wires inserted, forming a cage for the fruit. The toothed end of piece *b* is sharp, and slides over the end of *a,* which may be sharp or not. A small hole is bored through the pole, and a notch cut in the front edge for a small pulley, *d.* A strong cord is attached to the lower end of *b,* and passes through the hole over the pulley, and down the pole through screw-eyes placed a short distance apart. Figure 207 is a section of the lower end of the pole. Eighteen inches from the end, the pole is squared for about fifteen inches. Over this squared portion is fitted a sliding-box handle. A thumb-stop is fastened to the upper end, as shown in figure 208. The thumb end is held up by a small spring, which presses the upper end into notches in an iron rachet-bar fitted into the pole. A screw-eye is inserted in the upper end, and a cord attached. The pole may be of any desired length.

To pick apples, grasp the pole at the lower end with

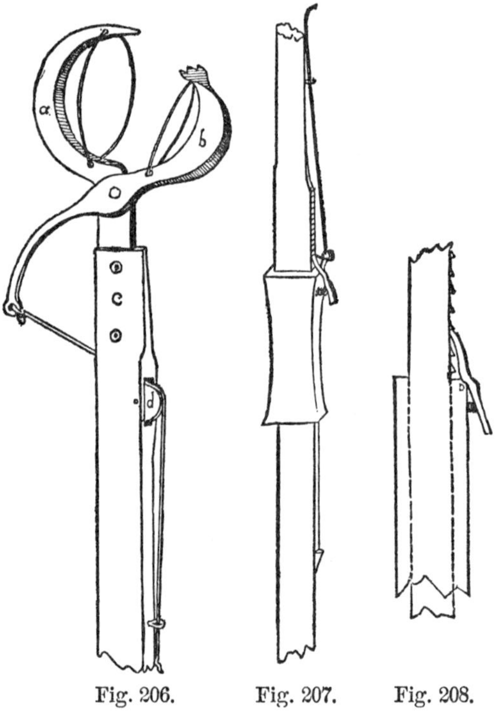

Fig. 206. Fig. 207. Fig. 208.

one hand, and by the sliding-box handle with the other. Press the thumb-piece and slide it up, and the weight of piece *b* opens the jaws of the picker. When the apple is in the cage, draw the slides down until the points of the picker meet on the apple stem. The thumb-stop will hold it secure. Turn the pole slowly without pulling, pushing, or shaking the limb, and the apple will come off easily. The cage of the picker should be large enough to contain the largest apple, and enough wires may be attached to hold the smallest. The jaws should not be over one-eighth of an inch thick, flattened on the inside, to prevent bruising the ripe fruit. They may be wrapped with cloth, if thought necessary.

A cheap and simple picker may be made by bending a

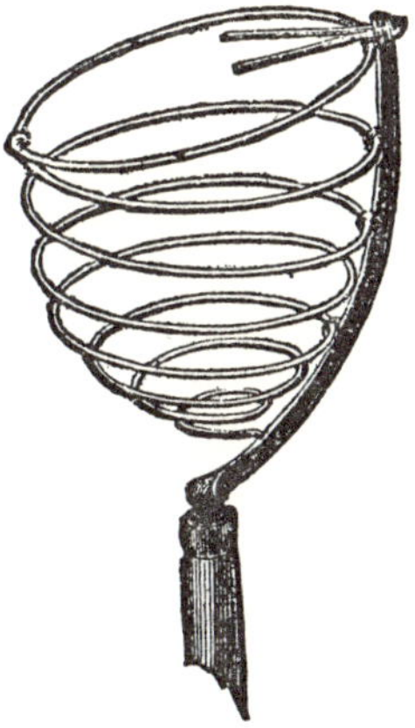

Fig. 209.

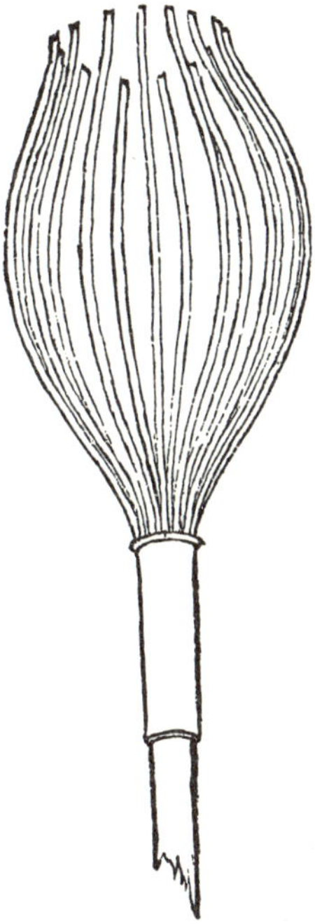

Fig. 210.

Fig. 211.

stiff wire into the form of a circle six inches in diameter, with one side of the circle prolonged three inches into a V-shaped projection. Upon this wire sew a cloth bag a foot or so deep, and fasten it on to a pole by the end opposite the V-shaped extremity. This V-shaped projection will serve as a corner, in which to catch the apple and pull it off, allowing it to fall into the bag. An excellent picker, as shown in figure 210, can be made from stiff wire by a tinner. The span across the top should be about six inches, and the depth from eight to ten inches. The wires should not be more than a half-inch apart at their tips. The wires being more or less flexible, the apple is apt to draw through them, if they are not close together. Care should also be taken to have the implement made as light as possible. A bungling mechanic will probably use too much solder. Another good picker is pictured in figure 209. It is light, durable and pleasant to handle. When, however, an apple, being very short stemmed, lies close to a limb, it is much more easily removed by the former device than by this. A simple, flattened hook, with a thin, almost cutting edge, secured on the end of a pole, figure 211, is often handy for pulling off stray apples. This is the best implement for thinning apples.

FRUIT LADDERS.

The construction is easily understood from the engraving. The method of using deep fruit baskets with a hook attached is also shown in figure 112. The use of a common grain bag as a receptacle for picking fruit has some important advantages. One side of the mouth of the bag is tied to the corresponding corner at the bottom, first putting an apple in the corner to hold the string from slipping off. The bag is then hung over the shoulder with the mouth in front. The picker has both hands free and can empty the bag by lowering it into the barrel, without bruising the fruit.

Another form is shown in figure 213. To make it, select a chestnut pole, eighteen feet long, or of the desired length. At about four feet from the top, or smaller end of the pole, nail on a band of hoop iron, to prevent splitting, and rip up the pole in the center as far as the band. The halves of the pole are spread apart three and a half feet at the base, and secured. The places for the rungs are then laid out, and the holes bored; those for the lower rungs should be one and three-eighths inch, the upper one inch; drive them in place and wedge fast.

Fig. 212.—A HANDY FRUIT LADDER.

Fig. 213.—FRUIT LADDER.

The distance between the rungs is usually a foot; when farther apart, they are fatiguing in use. A ladder of this kind, on account of its small width above, is easily thrust in among the branches, without breaking them, and is more convenient to use on large trees, than those of the ordinary shape.

JAPANESE PRUNING SAW.

The Japanese use a pull saw instead of a push saw. One of these is quite handy, especially for pruning. The teeth are like those of a rip saw, reversed, and cut when the saw is pulled towards one. One of these saws, made

Fig. 214.—PRUNING SAW.

as shown in figure 214, and fixed to a pole of convenient size, will be found very useful in cutting branches of tall trees, as in pulling there is no tendency to bend the saw or the pole.

RABBITS AND MICE IN THE ORCHARD.

Not the least of the enemies of young orchard trees is the rabbit. He will not injure the trees in summer, when he has an abundance of succulent food ; but in winter the tender bark is to him a dainty that he will partake of, if it is not made distasteful to him, or he is not kept away. Making the snow into a solid mound about the tree will keep away mice, but not rabbits, though it is often said it would. The rabbits will get on the mound and nibble away. Besides, we don't have snow half the time during the winter. The best way is to make the bark distasteful to the rabbit. He likes neither blood, nor grease, nor the odor of flesh. When you butcher, take the waste parts of the animals, and with these parts rub the trunks as far up as the rabbits can reach. The rabbits never nibble a tree so treated, while the grease or blood remains.

If the rabbits "bark" a tree, the first thing to be done is to examine the extent of the injury. Frequently it is not so bad as it looks, and the inner bark is not entirely removed. If this covers even a fourth of the wounded portion, and connects the bark above the wound with that below it, the chances are that the wound will heal, if drying can be prevented. The ordinary grafting wax, applied on old, worn cotton cloth, or on paper, as used in grafting, should be applied over the injured portion. This, especially on quite small trees, will prevent all evaporation. Another application is the old grafting clay, made by

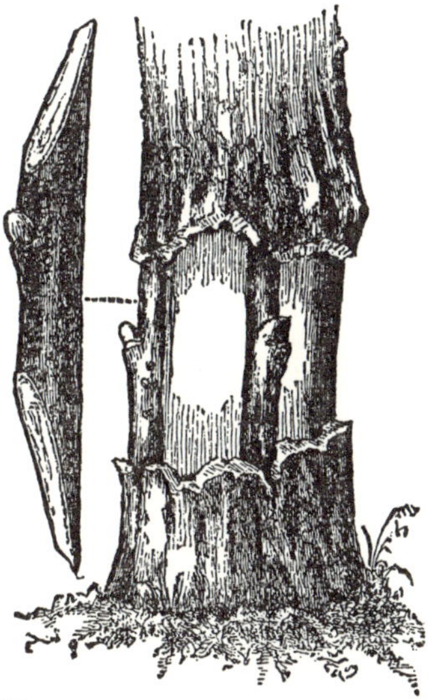

Fig. 215.—MANNER OF INSERTING THE CIONS.

thoroughly mixing and beating together stiff clay with half as much cow manure. Apply this over the wound quite thickly, and fasten it in place by wrapping with an old cloth and tying with strings. If the inner bark is completely gone, nothing remains but to bridge over the wound with cions, and thus restore the communication between the roots and top. The cions may be taken from the same tree, if they can be spared, or those from another of the same kind will answer as well. The methods of cutting the cions and inserting them are so plainly shown in figure 215 that description is unnecessary. A small chisel may be used to aid in setting the cions. This method of cutting the ends is better than making the slope on the opposite side. If the wound is low enough, it may be covered with a mound of earth; if not, employ one of the methods suggested above.

IMPLEMENTS USED IN CRANBERRY CULTURE.

A turfing axe, shown figure 216, consists of a thin steel blade, hatchet-faced, and about six inches square.

Fig. 216.—TURFING AXE.

This blade is made fast to a stout hickory handle, some two feet and a half long, in the same manner as a common wood axe. In expert hands, this axe does wonderful

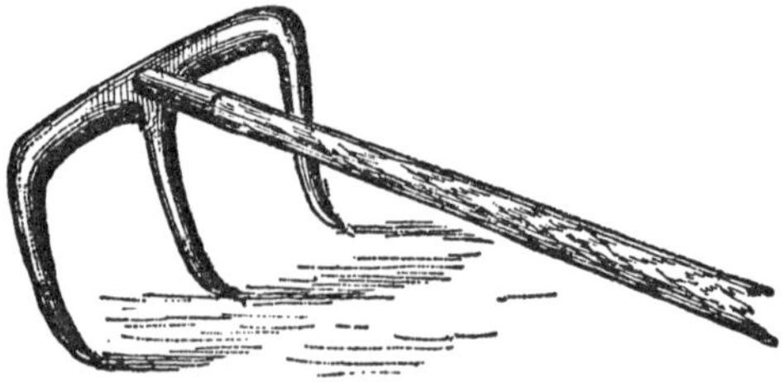

Fig. 217.—HAULING RAKE.

execution upon the tough, interlacing roots, with which the surface of the bog is filled.

A hoe, shaped like a grubbing hoe, is the implement used for grading. Every farmer knows what that is; but the grading hoe, figure 218, should be made of the best steel, and ground to an edge like an axe—the object being to cut all the fine roots to pieces, and get out such

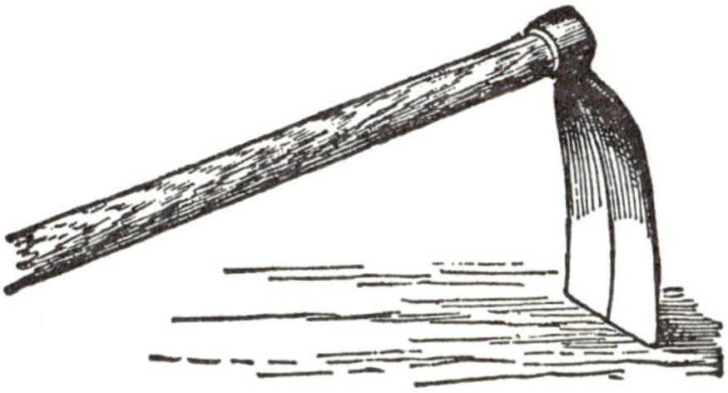

Fig. 218.—GRADING HOE.

of them as escaped when the trees, stumps, shoots, and larger wood were removed.

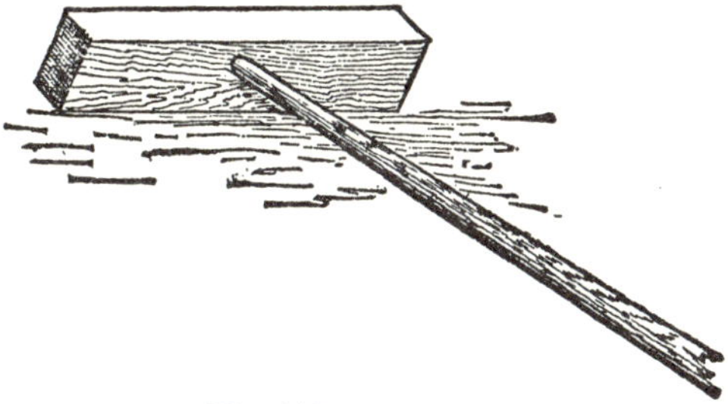

Fig. 219.—SPREADER.

The sand is spread by means of a "Spreader," figure 219, made of a piece of one-inch white oak board,

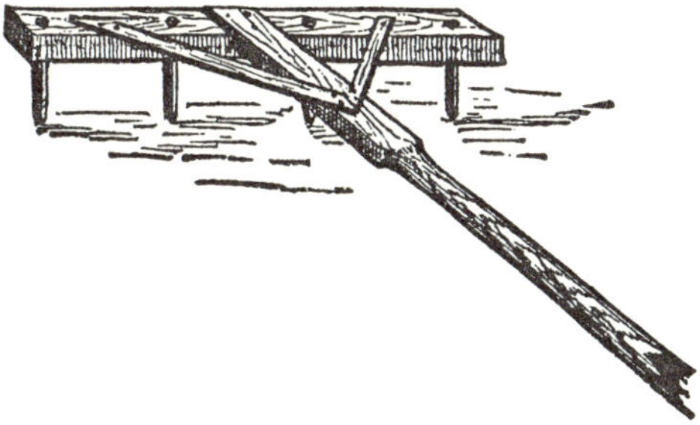

Fig. 220.—MARKER.

about fifteen inches long by three inches wide, and fastened to a handle.

A "Marker," shown in figure 220, is made of a piece of two by four inch joist, about nine feet long, having teeth eighteen inches apart, and a handle the length of a rake-handle. The teeth are eight inches long, made of white oak, driven through holes bored in the joist for the purpose. The implement is made similar to a common rake with teeth far apart, and the whole made stronger to stand harder usage, by having stays running from the handle to the head, which holds the teeth.

CHAPTER X.

APPLIANCES FOR SLAUGHTERING HOGS AND CURING THE MEAT.

STICKING HOGS.

The usual method of killing hogs on the farm is to thrust a sticking knife into the throat, severing the large veins. It requires experience, nerve, and skill to do this properly. The hog should be thrown on its back, and held there by an assistant, while the operator gives the fatal thrust. With a keen double-edged knife in his right hand, he feels with his left for the proper place to insert the knife. Having found it, he sticks in the knife, aiming directly toward the base of the tail. If properly done, the large veins are severed, and the hog soon bleeds to death. If the knife veers to either side, a gash is made in one shoulder, the death is slow and painful, and the blood settles in the flesh.

A BETTER WAY.

With a view to avoiding all mishaps, saving pain, and leaving the operator free to sever the veins without embarrassment from the squealing and struggling victim, the design called "The Stunner," figure 221 has been invented. It fits over the head of the intended victim, as seen in figures 222 and 223, and a sharp blow on the plate over the forehead drives the pin into the brain,

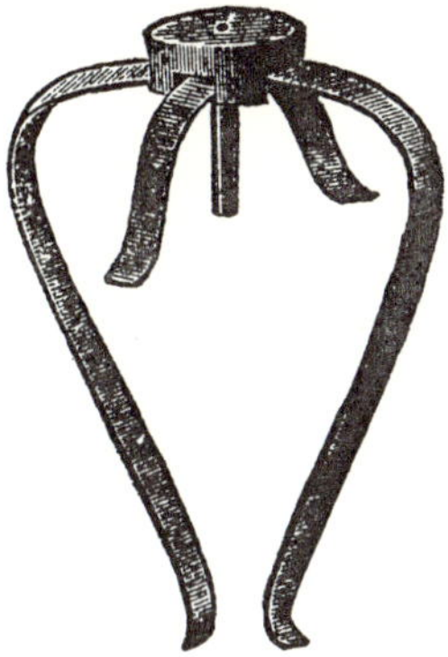

Fig. 221.
THE STUNNER.

Fig. 222.

Fig. 223.
THE STUNNER ADJUSTED.

causing insensibility instantly, and death will not be long delayed. The use of such a mask is made compulsory in many countries of Europe. Such a contrivance is not only convenient, but humane, and appeals to the better nature of every man who is under the necessity of killing a dumb beast. As soon as the animal is struck, the throat is cut to insure free bleeding.

HEATING THE WATER FOR SCALDING.

For heating scalding water and rendering lard, when one has not kettles or cauldrons ready to set in brick or stone, a simple method is to put down two forked stakes firmly, as shown in figure 224, lay in them a pole to support the kettles, and build a wood fire around them on the ground. A more elaborate arrangement is shown in figure 225, which serves not only to heat the water, but as a scalding tub as well. It is made of two-inch pine boards, six feet long, and two feet wide, rounded at the ends. A heavy plate of sheet iron is nailed with rod nails on the bottom and ends. Let the iron project

Fig. 224.—HEATING THE WATER.

about one inch on each side. The ends, being rounded, will prevent the fire from burning the wood-work. They also make it handier for dipping sheep, scalding hogs, or

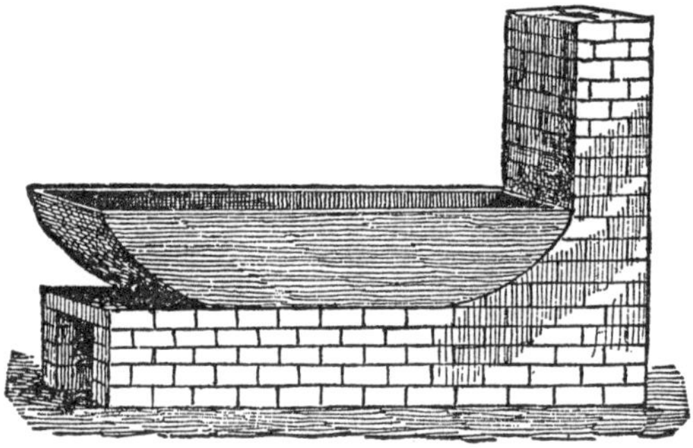

Fig. 225.—HEATING AND SCALDING VAT.

for taking out the boiled food. The box is set on two walls eighteen inches high, and the hind end of the brick-work is built into a short chimney.

SCALDING TUBS AND VATS.

Various devices are employed for scalding hogs, without lifting them by main force. For heavy hogs one may use three strong poles, fastened at the top with a log chain, which supports a simple tackle, figure 226. A very good arrangement is shown in figure 227. A sled is made firm with driven stakes, and covered with planks or boards. At the rear end the scalding cask is set in the ground, its upper edge on a level with the platform, and inclined as much as it can be and hold sufficient water. A large, long hog is scalded one end at a time. The more the cask is inclined, the easier will be the lifting.

Fig. 226.—TACKLE FOR HEAVY HOGS.

A modification of the above device is shown in figure 228. A lever is rigged like a well sweep, using a crotched stick for the post, and a strong pole for the sweep, a white oak stick—such as every farmer who can do so, should have laid up to season. The iron rod on which the sweep moves must be strong and stiff. A trace chain is attached to the upper end, and if the end of the chain has a ring instead of a hook, it will be quite convenient. In use, a table is improvised, unless a strong one for the purpose is at hand, and this is set near the barrel. A noose is made with the chain about the leg of the pig,

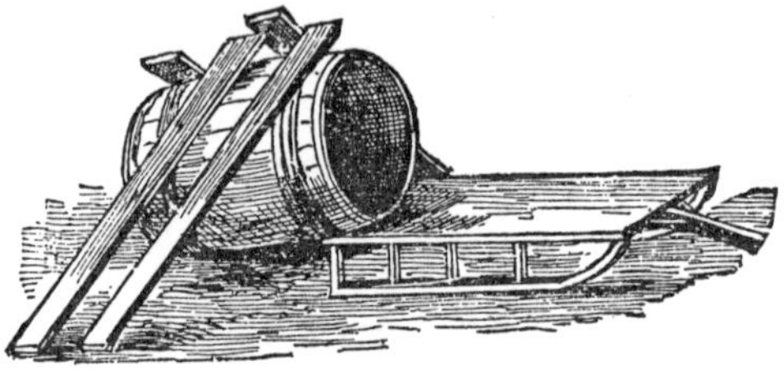

Fig. 227.—SCALDING CASK ON A SLED.

and he is soused in, going entirely under water, lifted out when the bristles start easily, and laid upon the table, while another is made ready.

Figure 229 shows a more permanent arrangement. It is a trough of plank, with a sheet iron bottom, which can be set over a temporary fire-place made in the ground.

Fig. 228.—SCALDING PIGS IN A HOGSHEAD.

The vat may be six feet long, three feet wide, and two and one-half feet deep, so as to be large enough for a good-sized hog. Three ropes are fastened on one side, for the purpose of rolling the hog over into the vat, and rolling it out on the other side when it is scalded. A number of slanting cross-pieces are fitted in, crossing each other, so as to form a hollow bed in which the carcass lies, with the ropes under it, by which it can be

Fig. 229.—SCALDING VAT.

moved and drawn out. These cross-pieces protect the sheet-iron bottom, and keep the carcass from resting upon it. A large, narrow fire-place is built up in the ground, with stoned sides, and the trough is set over it. A stove-pipe is fitted at one end, and room is made at the front by which wood may be supplied to the fire, to heat the water. A sloping table is fitted at one side for the purpose of rolling up the carcass, when too large to handle otherwise, by means of the rope previously mentioned. On the other side is a frame made of hollowed boards set on edge, upon which the hog is scraped and cleaned. The right temperature for scalding a hog is one hundred and eighty degrees; and with a thermometer there need be no fear of overscalding, or a failure from the lack of

Fig. 230.—A GAMBREL.

sufficient heat; while the water can be kept at the right temperature by regulating the fuel under the vat. If a spot of hair is obstinate, cover it with some of the removed hair, and dip on hot water. Always pull out hair and bristles, shaving any off leaves unpleasant stubs in the skin.

Gambrels should be provided of different lengths, if the hogs vary much in size, like figure 230, or in other convenient shapes. These should be of hickory or other tough wood, for safety, and to be so small as to require little gashing of the legs to receive them.

HANGING AND CLEANING THE HOGS.

Figure 231 shows a very cheap and convenient device for hanging either hogs or beeves. The device is in shape much like an old fashioned "saw-buck," with the lower rounds between the legs omitted. The legs, of which

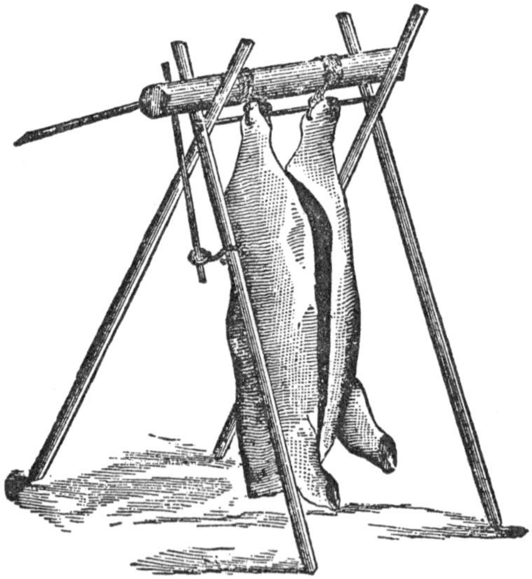

Fig. 231.—RAISING A SLAUGHTERED ANIMAL.

there are two pairs, should be about ten feet long, and set bracing, in the manner shown in the engraving. The two pairs of legs are held together by an inch iron rod, five or six feet in length, provided with threads at both ends. The whole is made secure by means of two pairs of nuts, which fasten the legs to the connecting iron rod. A straight and smooth wooden roller rests in the forks made by the crossing of the legs, and one end projects about sixteen inches. In this two auger holes are bored, in which levers may be inserted for turning the roller. The rope, by means of which the carcass is raised, passes over the roller in such a way that in turning, by means of the levers, the animal is raised free from the ground. When sufficiently elevated, the roller is fastened by one of the levers to the nearest leg.

Skill and practice are needed to take out the intestines neatly, without cutting or breaking them and soiling the flesh. Run the knife lightly down, marking the belly straight, cut to the bone between the thighs, and in front of the ribs and below, and split the rear bones with an axe carefully, not to cut beyond them ; open the abdomen by running the hand or two fingers behind the knife with its edge turned outward. Little use of the knife is required to loosen the entrails. The fingers, rightly used, will do most of the severing. Small strong strings, cut in proper lengths, should be always at hand to quickly tie the severed ends of any small intestines cut or broken by chance. An expert will catch the entire offal in a large tin pan or wooden vessel, holding it between himself and the hog. Unskilled operators, and those opening very large hogs, need an assistant to hold this. The entrails, and then the liver, heart, etc., being all removed, thoroughly rinse out any blood or filth that may have escaped inside. Spread the cut edges apart by inserting a short stick between them, to admit free circulation of cool air. When dripping is over, or the hanging posts are wanted for other carcasses, remove the dressed ones, and hang them in a cool cellar or other safe place, until the whole flesh is thoroughly cooled through. Removing the lard from the long intestines requires expertness that can only be learned by practice. The fingers do most of this cleaner, safer and better than a knife. A light feed the night before killing leaves the intestines less distended and less likely to be broken.

PACKING PORK.

Pack closely in the barrel, first rubbing salt well into all exposed ends of bones, and sprinkle well between each layer, using no brine until forty-eight hours after, and then let the brine be strong enough to bear an egg. After six weeks take out the hams and bacon and hang in the smoke-house. When warm weather brings danger of flies, smoke a week with hickory chips, avoiding heating the air much. If one has a dark, close smoke-house, as the writer has, the meat can hang in all the summer; otherwise pack in boxes, putting layers of sweet, dry hay between. Long experience has convinced me that this method of packing is preferable to packing in dry salt or ashes. Much lard is injured or spoiled by overheating and burning some portions; the smallest quantity scorched gives a bad flavor to the whole. A bucket of water in the rendering kettle prevents this, if the fire is kept from rising too high around the sides. The water is easily separated at the bottom, if not slowly evaporated off during the rendering. Cutting the leaf, etc., fine with a sharp hatchet or cleaver, facilitates the free extraction of the lard.

BOX FOR SALTING MEATS.

A trough made as shown at figure 232 is very handy for salting meats, such as hams, bacon and beef, for drying. It is made of any wood which will not flavor the meat; ash, spruce or hemlock plank, one and a half inch thick, being better than any others. A good size is four feet long, by two and a half wide, and one and a half

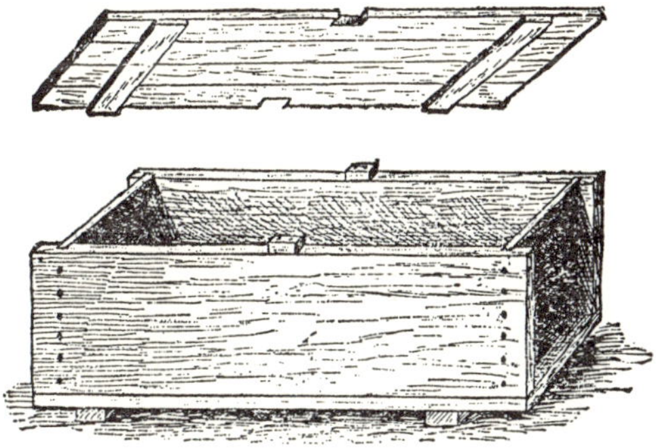

Fig. 232.—CONVENIENT SALTING TROUGH.

deep. The joints should be made tight with white-lead spread upon strips of cloth, and screws are vastly better than nails to hold the trough together.

MAKING SAUSAGES.

The quality will depend largely upon the kind of meat or meats used. Cutting fine is desirable, running twice or

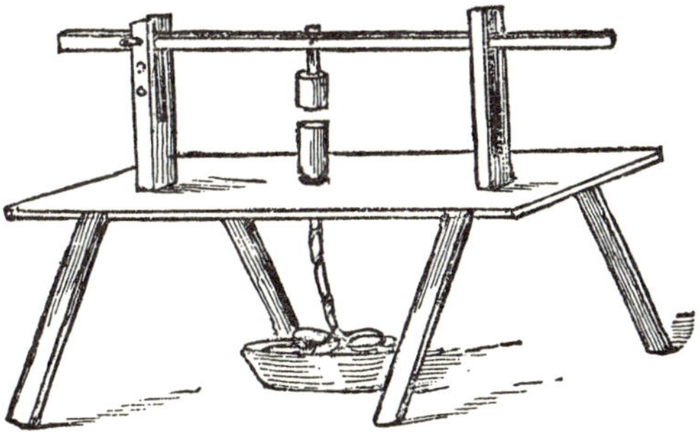

Fig. 233.—A SAUSAGE FILLER.

oftener through the machine, unless it be a better one than most of those in use. If skins are not used, the meat, after being cut and seasoned, may be pressed into cakes with the hands, the same as if wanted for frying, and packed in earthen jars. For long keeping, into summer if desired, make into suitable cakes and fry, pack in jars, and fill these with melted lard. The pieces can be taken out at any time and simply warmed through; they will be as sweet and fresh as when first prepared. Don't spoil sausage meat with spices or mint. Use salt and pepper moderately, leaving every one to apply these freely in eating, as individually desired. Figure 233 shows a simple bench and lever arrangement to be used with the common sausage filler, which lightens the work so much that even a small boy can use it with ease, and any person can get up the whole apparatus at home with little or no expense. An inch thick pine board, a foot wide, and four and a quarter feet long, is fitted with four legs, two and a half feet long, notched into its edges, with the feet spread outward to give firmness. Two oak standards, eighteen inches high, are set thirty-four inches apart, with a slot down the middle of each, for the admission of an oak lever eight feet long. The left upright has three or four holes above each other for the lever pin, as shown in the engraving. The tin filler is set into the bench nearer the left upright, and projects below for receiving the skins. Above the filler is a follower fitting closely into it, and its top working very loosely in the lever, to allow full play as it moves up and down. The engraving shows the parts and mode of working.

SMOKING THE MEAT.

The building shown at figure 234 is seven feet in length, six feet wide, and seven feet high from the ground floor to the lower side of the plate. It should be boarded up with matched stuff, or well battened. The frame is four by four timber at the top and bottom, to which the boards are nailed.

All smoke-houses should be fire-proof, but the expense involved in the construction of a brick or stone building, is more than many farmers and others care to incur. A wooden building, containing a fire-proof bin, that will not interfere with the use of the building for smoking meat, will answer. Any wooden smoke-house may be easily

Fig. 234.—IMPROVED SMOKE HOUSE.

and cheaply transformed to contain a fire-proof receptacle for ashes. In figure 234, is given a view of the interior of the smoke-house, showing plainly the form and location of the ash-bin. The bin is made from brick (long, narrow, flat stones will answer), and is built across one end or side of the building. The bricks are laid lengthwise, and one deep. The bin should be nearly three feet in height, and occupy about half the floor space of the building. For readily removing the ashes, coat the bottom and sides of the bin with mortar. No wood should be used. For smoking the meat, build the fire upon the earth floor close to the brick wall, suspend by four wires a piece of sheet-iron—a section of flattened stove-pipe will answer—two feet square, eighteen inches above the fire. The object of this is to spread the heat over a greater surface, and not allow the blaze to flash up to the meat. It also prevents the possibility of any pieces falling directly upon the fire. When not in use, this iron screen

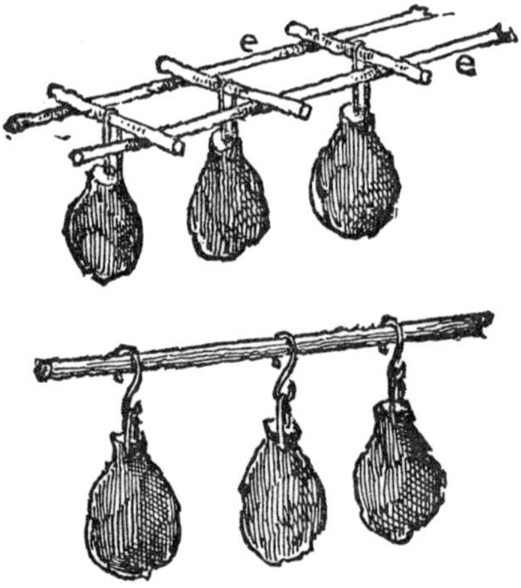

Fig. 235.—HANGING MEAT.

is unhooked from the poles above, and placed in one corner. Figure 235 shows the method of hanging the meat. As the meat must be carried from the cellar to the smoke-house, the short sticks make a secure handle for the purpose. The sticks rest upon cross-pieces, *e*, *e*, placed upon the plate of the building. Some prefer to use iron hooks, bent in the form of a large S, as shown in the lower part of the engraving, while by driving a spike or wooden pin in the edge or side of the scantling, a good support is obtained, figure 236. With a little more expense, a convenient support is obtained, so clearly shown in the upper part of the engraving, as to need no further description. In hanging up heavy pieces of meat, use strips of new or strong cloth, two inches in

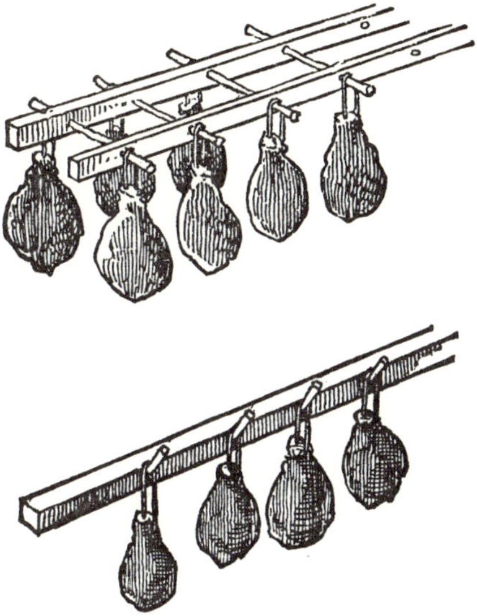

Fig. 236.—MEAT SUPPORTS.

width. A miniature wooden chimney should be placed at the top of the building, with an opening in the roof for ventilation, when required. Smoke from cobs or hickory wood imparts the best flavor to meat. The smoke-house door should, for convenience, open outward.

CHEAP TEMPORARY SMOKE-HOUSE.

For those who have only the hams and other meats from one or two hogs to smoke, a practicable smoke-house, like that shown in figure 237, will serve the purpose. A large barrel or good-sized cask should be used, with both heads removed. A hole about a foot deep is dug to receive it, and then a trench, of about the same depth and about six or eight feet long, leading to the fire-place. In this trench can be laid old stove pipe, and the ground filled in around it. What it is desired to have smoked is sus-

Fig. 237.—A BARREL SMOKE HOUSE.

pended in the barrel and the lid put on, but putting pieces under it, so there will be enough draught to draw the smoke through. By having the fire some distance from the meat, you get the desired amount of smoke, and avoid having the meat heated.

SELECTED BIBLIOGRAPHY.

Abler, William. *The Sensuous Gadgeteer.* Philadelphia: Running Press, 1973.

Familiarizes the reader with the basic working principles of various tools so that almost any simple project can be tackled with understanding. However it does not offer rural projects.

Blandford, Percy, W. *Old Farm Tools and Machinery, An Illustrated History.* Gale Press, 1976.

Bohlen, James. *The New Pioneer's Handbook: Getting Back to the Land in an Energy Scarce World.* Schocken, 1975.

Cohan, Ray. *How to Make It on the Land.* Englewood Cliffs, N.J.: Prentice-Hall, Inc., 1972.

Cook, Glen Charles. *380 Things to Make for Farm and Home.* Danville, Illinois: Interstate Printers and Publishers, 1941.

———. *500 More Things to Make for Farm and Home.* Interstate, 1944.

———. *600 More Things to Make for Farm and Home.* Interstate, 1951.

Together these three volumes comprise a virtual encyclopedia of plans for every conceivable kind of structure and equipment, with photos, drawings, and diagrams.

Davis, Fred. *Country Tools: Essential Hardware and Livery.* Oliver Press, 1974.

A good list of implements and their uses.

FITTING, MEL AND GREER. *In Praise of Simple Things.* New York: McKay, 1975.

HALSTED, BYRON D., ed. *Barns, Sheds and Outbuildings.* Brattleboro, Vermont: The Stephen Greene Press, 1977.

A practical manual offering a wealth of designs and plans for building utility structures of permanence and convenience. This is a facsimile edition of the book published by the O. Judd Co. in 1881 under the title *Barn Plans and Outbuildings.*

HARMOND, PAUL, ed. *The Farmstead Book, I.* Vancouver, B.C.: Cloudburst Press, 1978.

Excellent information on many build-it, do-it projects relevant to rural living.

JACKSON, ALBERT and DAY, DAVIS. *Tools and How to Use Them, An Illustrated Encyclopedia.* New York: Knopf, 1978.

Modern and traditional tools illustrated and explained.

JENKINS, J. G. *Traditional Country Craftsmen.* London: Routledge and Kegan Paul, Ltd., 1965.

Brickmaking, leather working, smithing, etc., illustrated and explained with the intent to inform about (not instruct in) the various arts.

KAYSING, WILLIAM. *First-Time Farmer's Guide.* San Francisco: Straight Arrow Books, 1971.

Advice on what the laid-back California farmer needs to know. In the milieu, but maybe not on target.

MARKS, VIC, ed. *Cloudburst.* Canada: Cloudburst Press, Ltd., 1976.

———. *Cloudburst II, A Handbook of Rural Skills*

and Technology. Canada: Cloudburst Press, Ltd., 1978.

Superb instruction in all manner of technological goodies: how to install, devise, make pumps, toilets, things, etc.

MARTIN, GEORGE A., ed. *Fences, Gates and Bridges.* Brattleboro, Vermont: The Stephen Greene Press, 1974.

A practical manual "showing the evolution of the fence from the road barrier of logs, brush or sods to the latest improved forms of barbed wire." So George Martin describes his book which covers every adjunct to fencing of concern to property owners, including wickets, gates, stiles, hedges, bridges, and culverts. This is a facsimile edition of the book published by the O. Judd Co. in 1887.

PARTRIDGE, MICHAEL. *Farm Tools Through the Ages.* Boston: New York Graphic Society, 1973.

Historical, illustrated—every kind of farm tool explored and explained according to its *historical* use.

ROBINSON, DAVID. *The Complete Homesteading Book, Proven Methods for Self-Sufficient Living.* Charlotte, Vermont: Garden Way Publishers, 1974.

Well-illustrated guide for everything from buying land to setting up systems for survival. Instructional.

SLOAN, ERIC. *Diary of an Early American Boy.* New York: Ballantine, 1977.

Diary form with notes, sketches and instructions on skills necessary for daily living in an old-time world.

TETRAULT, JEANNE and SHERRY THOMAS. *Country Women, a Handbook for the New Farmer.* New York: Doubleday/Anchor, 1975.

How to do it without much cash outlay. Designs for buildings, implements to make caring for animals, growing vegetables and building by oneself easier. Pumps, tools, garden ways, etc. explained.

WIGGINTON, ELIOT, ed. *The Foxfire Book.* New York: Doubleday/Anchor, 1969.

Always good instruction in with the tales and folklore. Complete log cabin instructions, thatching chairs, making soap are just a few of the how-to's.

WRIGHT, PHILIP, A. *Old Farm Implements.* Devon, U.K.: David and Charles, 1975.

INDEX.